A BRIEF INTRODUCTION TO

ASTRONOMY IN THE MIDDLE EAST

BRIEF INTRODUCTIONS SERIÉS

A BRIEF INTRODUCTION TO

ASTRONOMY IN THE MIDDLE EAST

JOHN M. STEELE

SAQI

London San Francisco Beirut

ISBN: 978-0-86356-428-4

First published by Saqi, 2008

A full CIP record for this book is available from the British Library.

A full CIP record for this book is available from the Library of Congress.

Manufactured in Lebanon

SAQI

26 Westbourne Grove, London W2 5RH
825 Page Street, Suite 203, Berkeley, California 94710
Tabet Building, Mneimneh Street, Hamra, Beirut
www.saqibooks.com

Contents

Illustrations

Glossary

acronychal rising The evening on which a star or planet is first visible above the eastern horizon after sunset.

altitude The angular distance of a body above the horizon.

anomaly An irregularity in the motion of a celestial body (or a phenomenon affected by its motion) caused by the variable velocity of the body. For example, the length of the lunar month is affected by *solar anomaly* and *lunar anomaly*.

azimuth The angular distance of a body around the horizon measured clockwise from north.

celestial equator The projection of the Earth's equator onto the celestial sphere. The daily rotation of the Earth is seen by an observer on the Earth's surface as a full circuit of the celestial equator in the sky.

celestial latitude or latitude The angular distance of a celestial body measured above or below (i.e., perpendicular to) the ecliptic.

celestial longitude or longitude The angular distance of a celestial body from a defined zero-point (usually 0° in Aries) measured parallel to the ecliptic.

conjunction | The instant when two celestial bodies have the same celestial longitude.

declination | The angular distance of a celestial body measured above or below (i.e., perpendicular to) the celestial equator. Celestial bodies with the same declination rise over the same point on the horizon.

deferent | In Greek geometrical astronomy the circle that carries the epicycle.

direct motion | The apparent motion of a planet with increasing celestial longitude.

eccentric | In Greek geometrical astronomy a circle whose centre is offset from the Earth.

ecliptic | The great circle defining the path of the sun.

ecliptical coordinates | A coordinate system using the ecliptic as its frame of reference. The two coordinates are *latitude* (measured perpendicular to the ecliptic) and *longitude* (measured parallel to the ecliptic).

elongation | The difference in longitude of two celestial bodies, often where one of these bodies is the sun.

epicycle | In Greek geometrical astronomy a secondary circle carried on the deferent circle used to generate variable motion.

equant | In Greek geometrical astronomy a device for producing non-uniform circular motion. The *equant point* is a point offset from the centre of a circle about which the motion appears uniform.

equatorial coordinates | A coordinate system using the equator as its frame of reference. The two coordinates are *declination* (measured perpendicular to the equator) and *right ascension* (measured parallel to the equator).

equatorial hours | One twenty-fourth part of the day, corresponding to one twenty-fourth part of the celestial equator.

equinox | The moments when the sun is at one of the two points of intersection of the ecliptic and the celestial

	equator. At the equinoxes the length of day and night are equal.
first appearance	The first day on which a celestial body is seen (just before sunrise in the case of an outer planet) after a period of invisibility caused by it being too close to the sun.
gnomon	A vertical pole whose shadow is used to observe the position of the sun.
inner planet	Mercury and Venus.
last visibility	The last day on which a celestial body can be seen (always just after sunset in the case of an outer planet).
node	The two points where the orbit of the moon crosses the ecliptic.
obliquity of the ecliptic	The angle at which the ecliptic intersects with the celestial equator.
opposition	The instant when two celestial bodies are separated by 180° in celestial longitude.
outer planet	Mars, Jupiter and Saturn.
precession	The gradual change in the point of intersection of the ecliptic and the celestial equator over time due to the slow change in the axis of the Earth's spin. The effect of precession is seen in the gradual increase of the longitudes of the fixed stars over time.
retrograde motion	The apparent motion of a planet with decreasing celestial longitude.
right ascension	The angular distance of a celestial body from a defined zero-point (usually 0° in Aries) measured parallel to the equator.
seasonal hour	One-twelfth of the period of daytime or night-time. Seasonal hours vary with the length of daylight over the seasons.
sidereal phenomena	Phenomena that depend upon the positional relationship between a celestial body and the fixed

stars, for example a conjunction or passage of a planet by a particular star.

solstice
: The two moments when the sun is furthest away from the celestial equator (i.e., at its maximum or minimum declination). At the winter solstice the length of daylight is shortest. At the summer solstice the length of daylight is longest.

stationary points
: The two points where a planet changes from direct to retrograde motion.

synodic arc
: The difference in celestial longitude between two successive synodic phenomena of the same kind.

synodic month
: The time interval between one conjunction of the sun and moon and their next conjunction.

synodic phenomena
: Phenomena that depend upon the positional relationship between a celestial body and the sun. Examples include the first and last visibilities, stationary points and acronychal risings of a planet and conjunctions and oppositions of the sun and moon.

synodic time
: The difference in time between two successive synodic phenomena of the same kind.

syzygy
: The moment when the sun and moon are in opposition or conjunction.

zenith
: The point with an altitude of 90° directly over an observer's head.

zodiac
: The division of the ecliptic into twelve equal parts each containing 30° of longitude. The twelve *zodiacal signs* are used as measurements of longitude.

Introduction

In 1068 AD a teacher and judge known as Sāʿid al-Andalusī decided to write a history of the scientific achievements of the nations of the world. Sāʿid lived in the Muslim-ruled city of Toledo in southern Spain, at that time a well-established centre of scientific learning. He called his book *Tābaqat al-umam* ('Book of the categories of nations'), and in it described the eight great nations that had developed science. He began with the ancient civilizations: India, Persia, Babylonia, Greece, Rome and Egypt. Each of these nations had contributed something important to the development of science. The Babylonians, for example, had gained great knowledge of the stars and their influence upon us. Of the Greeks, one man – Claudius Ptolemy – stood out as having reached perfection in the study of the heavens. Sāʿid next presented the current nations that he considered had contributed to science. Only two civilizations were worthy of inclusion: the Arabs and the Jews of Israel. The Jews excelled in medicine, but Sāʿid knew of no famous astronomers from Israel.

From what we know today about the history of astronomy we might object to Sāʿid's unjust neglect of China, which was named among his 'nations that have no interest in science', but was in fact a country with a long and impressive history of scientific achievement stretching back continuously to the first millennium BC. But otherwise Sā'id presented a remarkably accurate catalogue of the major cultures to have contributed to the history of astronomy. Only parts of the world that were unknown at the time were missing from his list: central and southern Africa, Australasia and the Americas, whose indigenous astronomies have only become known in Europe over the last few centuries.

Astronomy is, along with mathematics, one of the oldest sciences in the world. It is not always easy to remember today, in our age of streetlights and home entertainment, that the night sky seen on a dark night is a truly impressive and fascinating spectacle. Almost every culture on Earth has interacted with the sky. For thousands of years people have devised myths to explain the origin and appearance of stars and constellations. The motions of the sun and moon have been used to regulate everyday life through the development of calendars. Archaeological evidence from preliterate societies around the world has illustrated the importance of the sky to people. Even in regions such as Britain and Ireland, where the sky is often hidden from view behind clouds, prehistoric people incorporated the sky into their environment by aligning monuments and tombs in directions they considered important.

Two key developments were needed in order for astronomy to grow into a science. First came the means and desire to keep records of astronomical observations. Second was the realization that mathematics is a tool that could be used for astronomy. As far as we know, both developments first occurred in ancient Mesopotamia, although China was not far behind. The combination of regular astronomical records and the use of mathematics to study them allowed astronomy in Mesopotamia, principally in Babylonia, to flourish during the first millennium BC.

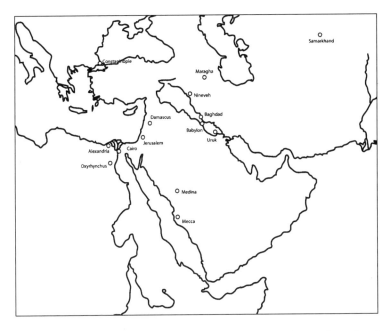

1. Map of the Middle East showing important ancient and medieval cities

The western astronomical tradition grew out of these Mesopotamian developments and was preserved and reformed by ancient Greek astronomers such as Hipparchus and Ptolemy. Their astronomy was in turn taken up by astronomers in the medieval Islamic world, who added their own innovations to the tradition. Eventually, this tradition would form the basis of the reform of astronomy in the European Renaissance by figures such as Nicholas Copernicus, Johannes Kepler and Galileo Galilei.

Astronomy has made remarkable advances over the past few decades. The development of larger, more powerful telescopes and new detectors has allowed astronomers to probe deeper and deeper into the universe, and to study celestial objects at ultraviolet and X-ray wavelengths beyond the visible spectrum of light. New theories within physics, such as quantum mechanics and Einstein's theories of relativity, and the power of computers for analysing astronomical observations and performing mathematical simulations of the evolution of the universe, have brought us closer to understanding the origin and growth of the universe in which we live. However, these exciting developments have often obscured the earlier history of astronomy from public view. Textbooks for introductory astronomy courses usually mention Kepler, Copernicus and perhaps Ptolemy, but nothing is said about the contribution of non-western cultures to the history of astronomical development. The result is that very few people outside the small academic world of history of early astronomy are aware that the astronomy that is practised today can trace its routes back to astronomers from many different cultures, religions and ethnicities, who spoke a variety of languages. The history of science is a multicultural story, a story that illustrates the links and similarities between civilizations rather than their differences.

Due to the dedicated work of a small group of groundbreaking scholars in the late nineteenth and twentieth centuries, we are now able to fill in many of the missing parts in the story of the development of astronomy from the first people who wrote down astronomical observations up to the present day. The rediscovery of Babylonian astronomy in the last part of the nineteenth century showed that it was in the Fertile Crescent in present-day Iraq that astronomy first arose. It was also in this region that astronomy was transformed into a mathematical science capable of making precise predictions of future astronomical phenomena.

Ancient Greek astronomers combined Babylonian astronomy with their own philosophical and geometrical approach to the movements of the heavenly bodies, producing mathematical models of the motions of the sun, moon and planets. In these models the Earth was at the centre of the universe and the planets moved around the Earth on complicated paths that resulted from combining circular motions. Claudius Ptolemy, one of the last and greatest astronomers of the ancient Greek world, developed planetary theories that were widely used throughout Europe and the Middle East until the sixteenth century.

Astronomers in the Islamic world took up the legacy of Greek astronomy during the medieval period. The Islamic faith brought astronomy into direct contact with wider society in a number of ways. The Islamic calendar, which defines the beginning of religious festivals such as Ramadan, is based upon the visibility of the new moon crescent at the beginning of each month. Astronomy also had other roles to play in religious practice. The times of the five daily prayers are defined by the position of the sun in the sky. Even finding the sacred direction towards Mecca, in which one should face when praying, was often treated as an astronomical problem in the medieval period.

Between the eighth and the fifteenth century, Islamic astronomers made many significant contributions to the development of astronomy. Perhaps the most visible example of their work was the production of astronomical instruments. Islamic instrument makers were innovators in the production of astrolabes, quadrants and other instruments. They developed new techniques for accurately building and marking scales on these instruments, and also excelled in producing instruments that were works of beauty, many of which are preserved in museums and private collections throughout the world. The instruments enabled astronomers to make precise observations of celestial phenomena, and to test and correct astronomical theories. The prevailing theories at that time were those of Ptolemy, but by the twelfth century astronomers were attempting to rectify what they saw as major flaws in Ptolemy's theories, using innovative mathematical techniques that were later taken up by astronomers during the European Renaissance.

With the decline of astronomy in the Islamic world at around the end of the fifteenth century, the locus of astronomy research switched to

Europe. Individual scientists such as Rogiomontanus, Nicholas Copernicus, Tycho Brahe and Johannes Kepler turned astronomy on its head. In the process they adopted and utilized several aspects of Islamic astronomy, and many Arabic terms entered the language of modern astronomy.

This book tells the story of the development of astronomy in the Middle East from the earliest times to the fifteenth century AD. The story begins in ancient Mesopotamia, continues in Greco-Roman Egypt and then in the medieval Islamic world that stretched from Persia, through the Middle East, along northern Africa and into Spain (see figure 1). During the fourteenth to seventeenth centuries, Arabic astronomy even reached as far as Timbuktu in Mali. We will encounter astronomical texts written in Akkadian cuneiform, Greek and Arabic, and astronomers who practised Islam, Christianity and Judaism, as well as a variety of ancient religions. All of these astronomers are part of our own astronomical tradition. Their energy, insightfulness and dedication have brought us to where we are today in understanding the universe that surrounds us.

The Birth of Astronomy
in the Middle East

The emergence of civilization in the Middle East took place more than 5,000 years ago in the region commonly known as Mesopotamia. The 'land between two rivers' (Mesopotamia) is the classical Greek name for the Fertile Crescent between the Rivers Tigris and Euphrates in what is now Iraq. By the end of the fourth millennium BC, several large cities had grown up in the southern Mesopotamian region, and the transition from small farming communities to large urban settlements brought with it a need for government and administration, which in turn led to the development of simple record keeping. Out of the practice of noting symbols on pieces of clay to denote the number of animals in a herd or the quantity of grain collected grew the world's first writing system. Know today as cuneiform, the script consisted of signs made up from wedges impressed in damp clay using a reed stylus. Each sign initially represented a word, but over time the script developed in such a way that a sign could also be used to indicate the phonetic sound of a syllable in a word from either the Sumerian or Akkadian languages that were used in the region.

Mesopotamia has a long and varied history of unification and insurrection, conquest and retreat. In the early part of the third millennium, several independent city-states ruled the neighbour-hoods around the major cities of southern Mesopotamia. Only occasionally did ambitious rulers attempt to bring a number of these city-states together, and none of them did this successfully over a long period of time. Despite the strong political independence of these city-states, trade and cultural interactions were frequent, and the Sumerian language was used throughout the area. The first successful unification of the Mesopotamian region took place when Sargon of Agade seized the throne of the state of Kish in a palace revolt and proceeded to conquer the other city-states one by one. Sargon's unification of Mesopotamia was to lead to profound changes in Mesopotamian history. Most importantly, it was during Sargon's time that Akkadian, a Semitic language related to Hebrew and Aramaic, became the main language throughout Mesopotamia, epitomized by it becoming the language used in official inscriptions.

Although Sargon's empire was to last for less than 200 years, the dominant feature of Mesopotamian history from his time onwards is large, unified political areas, rather than small individual city-states. While there were turbulent periods, successive rulers such as Hammurapi and Assurnasirpal II brought long periods of empire ruled from the city of Babylon in central Mesopotamia or the Assyrian heartland in northern Mesopotamia. In the first millennium Mesopotamia was finally conquered by outsiders: first by Cyrus the Persian who subsumed Mesopotamia into the Persian empire, and then by Alexander the Great who conquered the region for the Greeks. Despite the political upheavals these successive conquests brought in their wake, Mesopotamian culture, and especially Mesopotamian intellectual activity, continued apparently more or less unaffected. The latest dated cuneiform tablets were written in 75 AD. Interestingly, the three or four latest known tablets all concern astronomy. But let us begin much earlier, with the first appearance of the sun, moon, stars and planets in cuneiform texts.

The Earliest Evidence for Astronomy in Mesopotamia

The invention of writing in the fourth millennium BC acts as a cut-off date for gaining any detailed understanding of the thoughts and beliefs of the ancient Mesopotamians about the world around (and above) them. Before this time people had almost certainly looked at the sky, constructed patterns in the stars and noticed that some of the lights in the sky moved differently from others. Perhaps these moving stars were already associated with gods and stories composed to explain their origins and movements.

In the early part of the third millennium BC, very shortly after the invention of writing, a scribe in the southern Mesopotamian city of Uruk referred to a festival of the evening and morning god Inanna, who represents the planet Venus. The scribe who wrote this text evidently knew that the morning and evening stars were actually the same object, something that was not discovered in Europe for another 2,000 years.

For the ancient Mesopotamians, the heavens contained numerous stars and seven 'planets': the five planets visible to the naked eye (Mercury, Venus, Mars, Jupiter and Saturn), and the moon and the sun. Together these seven planets were called *bibbu*, which literally means 'wandering sheep'. By contrast, the stars and constellations were often compared with domesticated sheep, the sky being a cattle pen.

The Calendar

Before the development of writing, some form of basic calendar was probably already used by the people of Mesopotamia. A calendar fulfils as least two roles in most societies: it provides a means for bureaucracies to operate, to collect taxes and organize business transactions, and it acts as a framework for communal activities such as festivals and religious observances. In England, for example, the calendar provides days in the year both for celebrating Christmas or Halloween, and for making mortgage or rent payments or voting in elections. In some societies more than one calendar can be used at the same time: in the Muslim world, the date of the beginning of Ramadan is determined by the Islamic calendar based upon the visibility of the new moon, but everyone booking a flight from Tehran to London would use the western calendar, which is defined by the sun.

In Mesopotamia the calendar was based upon both the sun and the moon. Each month began on the evening when the thin crescent of the new moon was seen for the first time. If this took place on the thirtieth evening of the old month, the month was said to have been 'turned back' and that day instead became the first day of the new month. If the new moon crescent was not seen on the thirtieth day, the month was said to be 'completed' and the new month began on the following evening. This meant that months could have either twenty-nine or thirty days. On average there are almost the same number of twenty-nine-day months as thirty-day months over a period of a couple of years.

Twelve of these 'lunar' months make up about 354 days. This is about 11 days shorter than our year of 365 days (or 366 in a leap year). Our calendar is based upon the movement of the sun. In one of our years the Earth makes a complete circuit of the sun, or from our viewpoint on the Earth, the sun makes a complete path through the sky. As a result, our calendar stays in line with the seasons and we are guaranteed that our summer holidays will take place during the weeks when there is the best chance of good weather. In Mesopotamian calendars, however, a year containing 12 lunar months slips with respect to the seasons by 11 days every year. To overcome this problem the Mesopotamians introduced the concept of 'intercalary' months: extra months inserted into the calendar in some years in order to catch up to the seasons. These additional months, which have the same purpose as adding an extra day at the end of February in our leap years, were added roughly every three years. At first, these were inserted by royal command when the king accepted the advice from his advisers that an extra month was needed. This could lead to a very irregular system of inserting months, and some kings may have been reluctant to insert an extra month in a year if it meant delaying the collection of yearly taxes. In the second half of the first millennium the problem of irregular intercalation was avoided by the adoption of a scheme that assigned seven out of every nineteen years as intercalary years.

A calendar containing twenty-nine- and thirty-day months in no easily predictable order is not the easiest to use in administrative contexts. For example, in the calculation of interest on a loan, one must decide how to deal with months of different lengths. Should proportionally more interest be paid in thirty-day months than in twenty-nine-day months, or should the interest rate be treated monthly? In the latter case, how then do we deal with interest payments over periods of less than a month? And how will this be dealt with when the length of the month cannot be calculated in advance? To avoid this problem the Mesopotamians in the third millennium BC used an administrative calendar in which all months were taken to be thirty days long. Interestingly, a similar simplification of our calendar was used by financial institutions in the United States until the 1970s. In Mesopotamia this practise appears to have died out by the second millennium BC, at least in administrative contexts. This artificial calendar does, however, appear in many astronomical uses until the end of Mesopotamian scholarship.

In Mesopotamia the beginning of the new year was celebrated in the spring. The *akitu* festival began on the first day of the year and lasted for twelve days. It was by far the most important civil event of the year. Both the king and his subjects participated in the festival by performing specific cultic rituals on the different days.

Mesopotamian Views of the Universe

The Babylonian epic of creation, *Enūma Eliš*, which was probably composed towards the end of the second millennium BC, describes the creation of heaven and Earth. Marduk, the god of the city of Babylon and the hero of the tale, kills Tiamat, the goddess of the water who rose up to fight the other gods in revenge for her husband's murder, divides her body in two, and places one half to form the sky and the other to be the Earth. The epic continues:

> He fashioned stands for the great gods. As for the stars, he set up constellations corresponding to them. He designated the year and marked out its divisions, apportioned three stars each to the twelve months. When he had made plans of the days of the year, he founded the stand of Neberu to mark out their courses, so that none of them could go wrong or stray. ... He made the crescent moon appear, entrusted night (to it) and designated it the jewel of night to mark out the days. 'Go forth every month without fail in a corona, at the beginning of the month, to glow over the land. You shine with horns to mark out six days; on the seventh day the crown is half. The fifteenth day shall always be the mid-point, the half of each month. When Shamash looks at you from the horizon, gradually shed your visibility and begin to wane. Always bring the day of disappearance close to the path of Shamash, and on the thirtieth day, the [year] is always equalized, for Shamash is (responsible for) the year.

Although the epic is partly propaganda, establishing the primacy of the city of Babylon, the view of the universe as created by Marduk appears to reflect the thinking of the time. Marduk's universe is created not only complete with the stars and constellations, but also with an order and regularity in which the heavenly bodies are set in unchanging courses. The moon, for example, is assigned a thirty-day cycle of appearance, waxing, waning and disappearance. The year contains twelve months, each of which is associated with three stars. This basic view of the heavens is reflected in several other sources from the second and early first millennia. For example, a group of tablets named 'three stars each' in direct allusion to the statement in the epic of creation contains lists of stars in three paths for each month of the year. These paths, named after the gods Anu, Enlil and Ea, correspond to regions of the sky sweeping from the eastern to the western horizon.

Along with creating the heavenly bodies and setting them in motion, in this passage Marduk also implicitly creates the calendar. However, the calendar in *Enūma Eliš* is not that used in everyday life, but akin to the administrative calendar of the third millennium BC. Although the beginning of the month is defined by the new moon crescent, the month is always thirty days long and there are always twelve months in the year. The total number of days in a year is therefore 360, some six days more than in twelve true lunar months, yet still some five days less than in the solar year. This artificial 360-day calendar has therefore become known as the 'ideal calendar' by scholars working on the history of Mesopotamian astronomy: 'ideal' because the new moon does not always appear after thirty days.

The 360-day calendar appears in several other texts concerned with the heavens. The fourteenth tablet of the celestial omen series *Enūma Anu Enlil* contains four numerical tables that set out the duration of visibility of the moon and the length of daylight and night-time on dates in the ideal calendar. The first table illustrates their general form (figure 2). The first few lines read as follows:

The Moon			
	Day 1	3;45	It is present
	Day 2	7;30	It is present
	Day 3	15	It is present
	Day 4	30	It is present
	Day 5	1,0	It is present
	Day 6	1,12	It is present
	Day 7	1,24	It is present

For each of the thirty days during the month, the table gives the length of time the moon is visible during the night. These time intervals are written using the sexagesimal number system (see Appendix) in units of time called UŠ (pronounced 'ush'). There are 360 UŠ in a complete day and night, so 1 UŠ corresponds to 4 minutes of time. Thus on the first day the moon is present in the sky for 15 minutes, on the second day the moon is seen for 30 minutes, and so on. For the first five days of the month the length of time the moon is seen during the night doubles every day. For the following ten days this time interval increases by 12 UŠ (48 minutes) per day, until on the fifteenth day it will be visible all night for 180 UŠ (12 hours). This means that the data in the table relates to the two months containing the equinoxes during which night and day are of equal length. During the second half of the month the time interval decreases again following the opposite rule (see figure 3).

A second table on *Enūma Anu Enlil* tablet 14 gives essentially the same information for the length of visibility of the moon each night (see figure 3), albeit using different units of time. However, the two tables contain one important difference: in the second table the duration of the moon's visibility for the first and last five days of the month changes by 12 UŠ per day, rather than increasing and decreasing by a factor of 2. The motivation behind including these two alternate tables is not understood. The first table, for example, reflects the actual change in the length of visibility of the moon no better than the second. Indeed, the presence of these two tables seems to have troubled latter Mesopotamian scholars. A certain Bel-aba-usur, son of Bel-balāt-su-iqbī, a scribe from Babylon who lived in the middle of the first millennium BC, put together a text containing rules that rationalize the two systems. His first rule reads:

> [Day 1, the moon is visible for 3,45 UŠ (in the first table)] ... [The reciprocal of] 3,45 is 16. [16 multiplied by 4, the coefficient of visib]ility of the moon, [is 1],4. 1,4 multiplied by 3, the watch of your night, [is 3,12. 3,12 multiplied by 3,4]5 is 12. 12 multiplied by 1 is 12. Day 1 (the moon) is visible for 12 UŠ (in the second table).

But this is no more than a piece of sleight of hand, a mathematical

2. A Babylonian tablet containing a copy of the fourteenth tablet of the series Enūma Anu Enlil

trick akin to the child's game in which one 'predicts' the results of mathematical operations in which the beginning number has been kept secret by subtracting that beginning number somewhere during the process. In this case the author's sharp practice is to have us take the reciprocal of a number at the beginning, and then multiply by that same number later on. Expressing the rule using modern notation, if a is the length of time the moon is visible according to the first table, we simply have $1/a$ x 4 x 3 x a = 12 as the length of visibility according to the second table. The initial value simply drops out of the calculation, leaving us with an answer of 12 no matter what number we start with.

Similar mathematical schemes to those in *Enūma Anu Enlil* tablet 14 are found in a work called MUL.APIN. This is a two-tablet compendium of astronomical information including lists of stars and constellations, mathematical schemes giving the length of day and night throughout the year, and astrological omens. The day-length scheme also works within the framework of the schematic 360-day calendar. According to the scheme, the length of daylight at the summer solstice is twice that at the winter solstice. This 2:1 ratio for the length of day is a great exaggeration of the true variation in the length of day, more appropriate for the latitude of Paris than of central Mesopotamia.

The extreme inaccuracy of the 2:1 ratio for the length of day at the solstices leads to the question of what these early schemes were for. The use of the 360-day year raises the same question. Did the ancient Mesopotamians really think that the year contained 360 days and not know that the 2:1 ratio was grossly wrong? Or were these numbers simply nice round numbers that made the mathematics incorporated into these schemes very easy? Or did the schemes have another use that did not require accuracy? Two competing views have recently been put forward by historians.

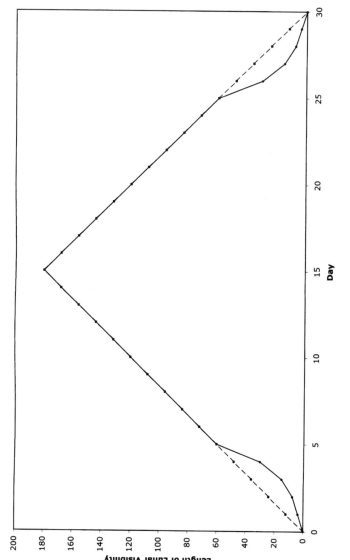

3. The variation in the duration of visibility of the moon during an equinoctial month according to the two schemes recorded on the 14th tablet of Enūma Anu Enlil

The first view interprets these schemes as basic approximations to nature. The 360-day schematic calendar provides a simple framework within which to make an initial calculation of an astronomical phenomenon, such as the length of visibility of the moon, which can then be corrected to the real lunar calendar. In support of this view, it is worth noting that although the 2:1 ratio for the variation in the length of daylight between the two solstices is very inaccurate, other variables derived from it, such as the daily change in the length of visibility of the moon, are fairly good approximations to the reality. These derived functions were probably of more importance to the Mesopotamians than the length of daylight itself. 'Why worry about an unimportant ratio's inaccuracy if it provides good results where it matters' may well have been the attitude.

The second view, recently proposed by D. Brown, argues that the mathematical schemes are not to be used for astronomical calculations. Instead they represent an ideal state for the universe against which observations can be judged. Discrepancies from the ideal were then interpreted as omens. In this understanding of the schemes, the ratio 2:1 and other numbers were chosen as much for their simplicity and perhaps magical significance as for their closeness to accurate values.

The Mesopotamian texts themselves provide scant evidence for any understanding of how we should interpret these schemes. What little evidence there is can be used in support of both of these two interpretations; quite possibly the schemes were used in both ways by the Mesopotamians.

Interpreting the Sky – Celestial Omens

The universe as understood by the Mesopotamians may have been created with an order and regularity, but it was also subject to the will of the gods. It was believed that the gods could, and did, send messages to humans. These messages took the form of 'signs' in the world around and above us. The signs could be interpreted by a diviner, who consulted texts containing lists of omens. Using his skills the diviner was then able to advise others on what the signs foretold and what action should be taken.

Two main forms of divination were practised in Mesopotamia. The first was the interpretation of omens that could be produced whenever they were needed through the action of the diviner. A modern fortune-teller may practise this form of divination by reading tea leaves or tarot cards. When a paying customer arrives, the reading of the omens can be initiated by the act of pouring out a cup of tea or dealing the cards. In ancient Mesopotamia the most common form of provoking an omen was through the sacrifice of a sheep and inspection of its entrails. Other, less gruesome provoked omens included interpreting the shape of oil as it spread after being dropped in water, or looking at shapes in the smoke from an incense burner.

The second type of omen interpreted by Mesopotamian diviners is drawn from uncommon experiences in the world. It is akin to the modern interpretation of a black cat crossing your path as a sign of bad luck, the finding of a four-leaf clover as a sign of good luck, and rain on St Swithun's Day forecasting forty more days of rain. In Mesopotamia unprovoked omens were taken from similar types of events: the entrance of wild animals into cities, anomalous animal births, human medical symptoms, weather changes, and occurrences in the day or night sky.

The earliest texts containing astrological omens date from the early part of the second millennium BC, but the compilation of several hundred omens into a canonical series seems to have taken place much later, around the end of the second millennium or the beginning of the first millennium BC. This series, known as *Enūma Anu Enlil* ('When the gods Anu and Enlil'), was written across seventy tablets and contained omens relating to the appearances of the sun and moon, eclipses, the appearances of the planets, the movements of planets through constellations, weather and more. All of the omens it contains are written using conditional if/then statements in the standard format used by the Mesopotamian diviners. For example, some of the Jupiter omens from *Enūma Anu Enlil* read as follows:

> If Jupiter passes at the right of Venus: Guti will be conquered with a strong weapon.

> If Jupiter passes at the left of Venus: Amurru will be conquered with a strong weapon.

> If Jupiter comes close to Venus: people will be thrown into confusion and brother will eat brother.

The apodosis ('then' statement) of the astrological omens always concerns the land as a whole. Only the king is mentioned as an individual since his actions and fate are tied up with those of the state. Personal astrology only appears in Mesopotamia during the second half of the first millennium BC (see the next chapter).

Many of the omens found in *Enūma Anu Enlil* refer to astronomical events that could not possibly take place. For example, we find omens concerning the appearance of Venus with different colours: red, black, white, green, and green and black. The same sequence of colours is found in other omen collections, both celestial and terrestrial, indicating that these Venus omens were created by following established procedures rather than as a result of past experience in noting astronomical observations and simultaneously occurring events on Earth. Similarly, *Enūma Anu Enlil* contains omens relating to the passage of Jupiter through the various parts of the constellation of the Scorpion: the front, rear, right, left, navel, sting and throat of the scorpion. It also applies the same attributes to the other constellations: the Goat, for example, is likewise assumed to possess a front, rear, right, left, navel, sting and throat. The celestial omens were created through a process of systematic extrapolation from known celestial phenomena to all imaginable astronomical events.

The Mesopotamian diviners did not believe that events in the heavens caused things to happen here on Earth. The omens were viewed only as signs, indications from the gods of what was to happen. A sign was like a warning light in a car indicating that it is low on fuel: the light does not cause the car to stop – it stops because it runs out of petrol. And just as a trip to the petrol station will prevent the warning coming true, the signs in the sky interpreted by the Mesopotamian diviners could be negated if the appropriate action was taken. A ritual called a *namburbû* (literally meaning 'its loosening') could be performed to make the evil portended by the omen pass by. Very few texts describing these *namburbû* rituals for astrological omens are preserved, but we can get an idea of the type of ritual from the *namburbû* for other types of omen. For example, if a pot squeaked in a man's house, the man should go to the side of a river and purify the site. He must then fill a water vessel with water from the temple of Marduk mixed with herbs and beads of metal and say a prayer to the sun-god Šamaš. Many reports of the kings from the Neo-Assyrian period (seventh century BC) performing *namburbû* rituals are preserved in official court documents.

One of the most ominous of astronomical events was an eclipse of the moon. Eclipses often foretold wars and battles, plague and famine, and even the death of the king. Such was their importance that there existed a special ritual performed at the occurrence of especially ominous eclipses. During the ritual, called *šar pûhi* (literally 'substitute king'), the king was replaced on the throne by a substitute in order that the evil foretold by the eclipse would befall the substitute instead of the true king. Whenever the king's advisers convinced him that the ritual was necessary, which was apparently quite often during the seventh century BC at least, a suitable substitute was chosen. The substitute would usually be a convicted criminal, a prisoner of war or a person at the very bottom of the social structure about whom the king and his court would not care very much. The king and the substitute would then officially switch roles: the substitute would become king and have to recite the omen associated with the eclipse. In this way he would accept responsibility for the king's shortcomings. The king, meanwhile, remained within the palace without any of the trappings of office, addressed only as 'the farmer', ruling from behind the scenes.

The substitute king ritual could last for up to 100 days before it was deemed that the evil had passed. During this time the substitute enjoyed most of the pleasures of the palace. He was looked after by a special entourage of cooks, musicians and others devoted to providing for the king's enjoyment. However, after the end of the ritual the substitute, in the role of the scapegoat, 'went to his fate' and was executed. He might then be given a royal burial. Meanwhile the true king would return to his throne, perhaps chastened by the experience, but otherwise unharmed by the evil portended by the eclipse.

Astrology in Practice – the Neo-Assyrian Period

The interpretation of astrological omens in ancient Mesopotamia, like the reading of astrological horoscopes today, depended as much upon the skills and motivations of the diviner as it did on the texts containing the compilations of omens. We have a unique opportunity to see this in action through the discovery of a large body of correspondence between the seventh-century BC Neo-Assyrian kings Esarhaddon and Assurbanipal and the group of scholars they employed to provide advice on the interpretation of omens. These texts were excavated by A. H. Layard, G. Smith and H. Rassam on behalf of the British Museum from the site of the Assyrian capital Nineveh in northern Mesopotamia. Many of the tablets were found in the South West Palace built by Esarhaddon's father Sennacherib at the end of the eighth century BC.

We only possess one side of the correspondence between the king and his scholars, namely the letters and reports sent by the scholars throughout the empire to the king. Many of the letters were written in answer to specific questions that the king posed about the interpretation of certain astronomical observations. For example, a certain Nergal-šumu-iddina wrote to the king as follows:

> To the king, my lord: your servant Nergal-šumu-iddina. Good health to the king, my lord! May Nabû and Marduk bless the king, my lord! Concerning the watch about which the king, my lord, wrote to me, the moon let the eclipse pass by, [it did not occ]ur. May [Sin and Šamaš appoint a gu[ardian of he]alth and life for the king, my lord.

Where an ominous event in the sky had been seen, the scholars would look up the interpretation of the omen in the omen series *Enūma Anu Enlil*, or sometimes take omens from unspecified other sources including oral tradition. They might also recommend to the king that a *namburbû* ritual should be performed. Since any particular astronomical event could usually be interpreted in many ways, the diviner often had quite considerable freedom to change the emphasis within the interpretation. For instance, if on a particular evening Jupiter first became visible near the head of the lion, and that evening also happened to be the first day of the month and the new moon rose in a halo, the diviner had at least four different pieces of evidence that could be interpreted ominously: the first appearance of the planet Jupiter, the position of Jupiter near a particular star, the date of the first visibility of the moon and the moon rising with a halo. The diviner could play off each of these items, together with combinations of them, against one another.

At any one time the king employed several scholars to advise him. In doing so he prevented any one scholar from gaining too much influence. One scholar alone could easily manipulate the interpretation of the omens for his own political or personal reasons, but a group of scholars provided a cross-check on their observations and recommendations. This is not to say that the scholars always agreed – far from it, in fact. The picture that emerges from the correspondence is one of competing individuals, each vying for the king's attention and favour. Indeed some of the scholars rubbished their competitors, informing the king that so and so could not have seen Venus on a certain night, or that such and such a person's interpretation of an astronomical phenomenon was incorrect. Sometimes a scholar would fall out of favour with the king. There is a long and maudlin letter of complaint from one scholar, asking the king why he has been forgotten and bemoaning his social state. 'People pass my house, the mighty on palanquins, the assistants in carts, (even) the juniors on mules, and I have to walk!' he complains. We do not know the king's response.

Among the astronomical observations reported in the Neo-Assyrian correspondence were the day the new moon crescent first became visible, which was of calendrical as well as ominous significance, the passages of planets through constellations, and the appearances of comets and meteors. By far the most important events, however, were eclipses. As described above, eclipses of the moon in particular were highly ominous, sometimes requiring the extreme measure of performing the substitute king ritual. Perhaps in order to allow for preparations to be made for this ritual, some of the Neo-Assyrian scholars started to try to predict eclipses in advance. The methods they used were based upon studying past occurrences of eclipses and noting certain regularities. For example, it was known that lunar and solar eclipses often occur within half a month of each other, and that when an eclipse was seen it would usually be another six months before another eclipse might be possible. In the next chapter we see how this first understanding of eclipses led to the development of a highly advanced theoretical approach to understanding the motions of the heavenly bodies.

Late Babylonian Astronomy

Throughout the latter part of the first millennium BC, the cultural and political locus within Mesopotamia shifted to the city of Babylon. Following a hundred years of uneasy dominance from Assyria punctuated by short periods of devolved rule or open rebellion, the Babylonian Nabopolassar seized the Babylonian throne in 626 BC and stood in open confrontation with Assyria. In the aftermath of Assyria's downfall in 612 BC, Babylonia reasserted itself as the dominant political power in Mesopotamia. However, Mesopotamia soon fell under the eye of neighbouring empires and in 539 BC the Persian king Cyrus marched on Babylon, only to be welcomed by the Babylonian people as a liberator from their unpopular king Nabonidus. Persian rule was to last for two centuries before Alexander the Great conquered Babylonia. Following internal squabbles after Alexander's death in 323 BC, his empire was split between Seleucus and Ptolemy. Seleucus assumed control of Mesopotamia and his dynasty ruled the region until it gradually fell to the Parthians in the second century BC.

The many changes of rule in Babylonia from around 750 BC to the first century AD, a period known as 'Late Babylonian', seems to have had little impact upon indigenous Babylonian culture, at least as far as the scribes who wrote in cuneiform were concerned. While in the general population Aramaic was fast becoming the common tongue, among the scribal elite Akkadian continued to be the language of authority, and cuneiform texts the medium in which information was preserved.

During much of the Late Babylonian period, it seems that most of Babylonian scholarship took place within the temples. We know very little about any of the individual astronomers from this time, other than a few names of scribes who owned or copied astronomical tablets. Some of them were employed in temples. Archaeological evidence also points towards the temples as the home of scholarship: many scholarly texts have been recovered from the sites of the temples in Uruk and Babylon.

The Late Babylonian period saw huge developments in the practice of astronomy: the first systematic recording of daily astronomical observations; the invention of a reference system, the zodiac, for measuring positions in the sky; and the formulation of mathematical methods for calculating astronomical phenomena. These are just some of the innovations in astronomy that came about in Babylonia during the first millennium BC.

'Regular Watching' of the Sky

There exists a group of astronomical cuneiform texts called by the ancient scribes EN.NUN *šá ginê*, literally 'regular watching'. As their name suggests, these texts contain records of night-by-night observations made by the Babylonian astronomers. For this reason these texts have been dubbed 'Astronomical Diaries' by modern scholars. More than a thousand fragments of Astronomical Diaries are known, making up about a quarter of all known Babylonian astronomical texts.

The earliest preserved Astronomical Diary contains observations made during the sixteenth year of Šamaššumukin (652–651 BC), and only a handful of these Diaries are known from before the fourth century BC. Nevertheless, it is almost certain that the Diaries were produced every year from at least 652 BC, perhaps as early as *circa* 750 BC, until the end of cuneiform scholarship in the first century AD. In other words, there was a continuous practice of recording nightly systematic astronomical observations for about 800 years. Compare this with modern astronomical observatories. Charles II founded the Royal Greenwich Observatory in 1675 AD, only a little over 300 years ago. The Vatican Observatory, founded up by Pope Gregory XIII to help reform the calendar, is slightly older. It dates to 1582 AD, over 400 years ago. But this is still well shy of the Babylonian tradition of more than 800 years.

Almost from the very beginning the Babylonians had a set format for what they observed and what they recorded in the Diaries. Each Diary typically covers half a year and is divided into sections for each month. A section begins with a statement of whether the new moon crescent was seen on the thirtieth or thirty-first evening of the previous month, and how long the moon was visible for on that night. I quote below from the Astronomical Diary containing observations from the thirty-seventh year of Nebukadnezar (568–567 BC) (see figure 4):

> Month XI, (the 1st of which was identical with) the 30th (of the preceding month), the moon became visible in the Swallow; sunset to moonset: 14° 30'; the north wind blew. At that time, Jupiter was 1 cubit behind the elbow of Sagittarius.

On this occasion the new moon was first seen on the thirtieth evening of the tenth month, so that the thirtieth day became instead the first day of the eleventh month (corresponding to 12 February 567 BC). The moon was seen for 14 UŠ (translated °) and 30 sixtieths of an UŠ (translated °), or 58 minutes of time, before it set. Also on this day Jupiter was observed to be 1 cubit behind a star known as 'the elbow of Sagittarius' (probably one of ξ, o or π Sagittarii). This distance is

4. The Babylonian Astronomical Diary for 568 BC

measured roughly parallel to the line of the ecliptic or the direction of the planet's motion. A cubit corresponded to approximately 2° of arc.

For the remainder of the month the Diaries give night-by-night reports of observations unless bad weather prevented any observations being made. The moon is most frequently recorded. Every evening the passage of the moon by one of a group of bright stars is recorded. These stars are known today as 'Normal Stars'. Twenty-eight stars are regularly cited and a few more are used occasionally. Other lunar phenomena recorded in the Diaries include six time intervals known as the 'lunar six'. The first 'lunar six' has already been met: it is the interval between sunset and moonset on the first day of the month. Four similar intervals are measured on the days around full moon, and the final interval is the time between moonrise and sunrise on the day when the moon was visible for the last time before its conjunction with the sun.

Eclipses of the sun and moon are one of the more impressive astronomical events recorded in the Diaries. Practically every observable eclipse was recorded by the Babylonian astronomers, although only a small fraction of these records is preserved today. The Babylonians achieved the consistent spotting of even very small eclipses by developing methods to predict at which new and full moons an eclipse might occur, and at what time of the day or night it would be likely to begin (see below). One of the most detailed eclipse accounts concerns a total solar eclipse that was seen in Babylon on 15 April 136 BC. The description of the observation is partially preserved on two tablets. One tablet only gives a brief account, stating that the eclipse began at 24 UŠ (= 1 hour 36 minutes) after sunrise, and after a further 18 UŠ (= 1 hour 12 minutes) it became total. The other account, however, is much more detailed:

> The 29th, at 24° after sunrise, solar eclipse; when it began on the south and west side, [... Ven]us, Mercury, and the Normal Stars were visible; Jupiter and Mars, which were in their periods of invisibility, were visible in its eclipse. [...] it threw off (the shadow) from west and south to north and east; 35° onset, maximal phase, and clearing.

Although the statement of totality is missing – it was presumably written in the broken part of the text – a clear description of the visibility of the planets and stars is given. This account is the most detailed description of a total solar eclipse known from the ancient world.

Observations of the planets are also recorded in the Diaries. These observations fall into two categories. First, there are reports of the days on which the planets passed above or below one of the Normal

Stars. These observations are known as 'sidereal phenomena', since they relate a planet to the fixed background of stars. The planets are not visible every night, however. Because both they and the Earth move around the sun, there will be periods when, viewed from the Earth, a planet will be too close to the sun to be seen. The planet will be in the sky during the day, but the light from the sun is too bright to allow us to see the planet. As the planet and the Earth move on their orbits, the planet will gradually move away from the sun until a critical day when it will be seen in the night sky for the first time shortly before sunrise. As the orbits continue the planet will rise earlier and earlier each night, moving through the stars. Then, one night, the movement of the planet will come to a stop and the planet will begin to move backwards ('in retrograde') through the stars. The point marking the change in direction of the planet is called its 'stationary point'. The planet will now continue to move in retrograde until it is visible all night, rising as the sun sets. After a further period, the planet will change direction once again, moving forwards once more through the stars, setting earlier and earlier each night until it comes too close to the sun and cannot be seen once more. This cycle, known as a planet's 'synodic cycle', applies to the three outer planets, Mars, Jupiter and Saturn. Because Venus and Mercury are closer to the sun than the Earth, these planets look from the Earth as if they are tied to the sun on a piece of elastic, never moving more than a certain distance away before moving back towards the sun. As a result, Venus and Mercury are only ever visible near sunrise and sunset, and appear as both morning and evening stars with two periods of visibility. The phenomena of the planets during their synodic cycle are called 'synodic phenomena'.

In addition to the planetary and lunar observations, the Diaries record some unusual astronomical phenomena such as comets and meteors. However, this type of observation is rare. It seems that the Babylonians were interested not only in regularly watching the sky, but also predominantly in watching regular phenomena. These are the phenomena that could be analysed and hopefully predicted.

One other phenomenon in the sky is regularly reported in the Diaries: the weather. Whenever clouds, mist or rain prevented an observation from being made, that fact was recorded in the Diary. Good weather and clear conditions were apparently not considered worth reporting.

Very little is known about how the Babylonians made their observations. Did they use instruments? Where did they stand to observe? Were the observations made by teams or by individuals? Unfortunately, on these issues the texts are silent. Nowhere in the Astronomical Diaries is there any mention of an observing instrument. From other sources we know that water clocks were used to measure time, but no examples have been found in the excavations of the Babylonian cities. It is quite possible that the only observing instruments used were the water clock, the naked eye and possibly something like a graduated stick held at arms' length to measure distances in the heavens.

The Invention of the Zodiac

Probably everyone today knows their 'star sign', the sign of the zodiac under which they were born: Aries, Taurus and the rest. If not we can look at the horoscopes in a daily newspaper and it will tell us the star sign for our birthday by listing a range of dates for each sign. These dates relate to the days during the year in which the sun is located within a particular zodiacal sign. Formally, however, the zodiac is defined by position not by date. Although modern astrology may claim possession of the zodiac (modern astronomers rarely use it any more), its origins go back to the astronomers of ancient Babylonia.

Sometime during the fifth century BC, the astronomers who compiled the Babylonian Astronomical Diaries developed a new method for recording planetary positions to go alongside their use of bright stars as a reference system. Instead of measuring the distance of a planet from a star, this new system was defined by the band of the heavens through which the sun, moon and planets move.

In MUL.APIN, the early two-tablet compendium of texts concerning the sky, there is a list of eighteen constellations through which the moon, sun and planets pass. These constellations are uneven in size and shape and therefore not particularly useful for tracking a planet's motion. The Babylonians who were interested in developing astronomical theories needed something better. So, in analogy with the 'ideal' year of twelve 30-day months totalling 360 days, they divided the zodiacal band into twelve equal strips and each of the twelve sectors into 30 UŠ or 'degrees', making a total of 360°. They named each strip after one of the constellations in that part of the sky. These strips became our signs of the zodiac.

Zodiacal Sign	Degree Range	Babylonian Name	Translation
Aries	0–30	HUN	The Hired Man
		(LU in some late texts)	
Taurus	30–60	MÚL-MÚL	The Stars (Pleiades)
Gemini	60–90	MAŠ-MAŠ	The Twins
Cancer	90–120	ALLA	The Crab
Leo	120–150	A	The Lion
Virgo	150–180	ABSIN	The Barleystalk
Libra	180–210	RÍN	The Balance
Scorpio	210–240	GÍR-TAB	The Scorpion
Sagittarius	240–270	PA	Pabilsag
Capricorn	270–300	MÁŠ	The Goat-Fish
Aquarius	300–330	GU	The Great One

Zodiacal Sign	Degree Range	Babylonian Name	Translation
Pisces	330–360	*zib*.ME	The Tails

Since the zodiac is an invisible reference system, the Babylonians needed to tie it down to something observable. Modern astronomy, following ancient Greek practice, uses the position of the sun on the ecliptic on a certain date (the spring equinox when day and night are of equal length) to fix the zodiac in the sky, but the Babylonians instead chose to map the zodiac onto their system of Normal Stars. For example, they placed the Rear Twin Star (β Geminorum) at the beginning of Cancer and the Rear Star of the Goat-Fish (δ Capricorni) at the beginning of Aquarius. Two fragmentary catalogues of Normal Stars giving their positions within zodiacal signs are known. These catalogues allowed the Babylonian astronomers to relate distances from Normal Stars with positions in the zodiac.

The development of the zodiac had a massive impact on the development not only of Babylonian astronomy, but also of ancient and medieval astronomy in the western world after the concept was adopted by the ancient Greeks. In Babylonia it allowed the creation of mathematical methods for calculating planetary and lunar positions, triggering a new phase in the science of astronomy. Astronomy was not the only beneficiary of the creation of the zodiac, however.

Horoscopic Astrology

Hand in hand with the invention of the zodiac as a frame of reference in the sky came the development of a new form of interpreting the movement and appearances of the heavenly bodies as forecasters of what was to happen on Earth. Whereas the predictions of traditional Mesopotamian celestial divination, epitomized by the omens in the series *Enūma Anu Enlil*, were directed towards the king and his rule over the country, this new form of astrology read the future of individuals in the sky. Key within this process was the study of the zodiacal signs in which the sun, moon and planets were situated at the time of an individual's birth. This practice of celestial birth forecasts has become known as 'horoscopic astrology' (although this term properly refers only to the Greek practice of calculating the *horoscopos*, the rising point of the ecliptic at the time of birth), and the texts where the astronomical data at the time of birth were recorded as 'Horoscopes'.

Only a handful of tablets with Horoscopes on them are known – twenty-eight in total from the cities of Babylon, Uruk and Nippur – perhaps because the Horoscopes would be purchased by individuals and taken to their homes, rather than being preserved in an archive. One can imagine that the clay from some Horoscopes might be reused once their owner was dead since the document would have been of little use to anyone else.

A typical Horoscope contained positions of the sun, moon and visible planets, either by zodiacal sign or more precisely in degrees within zodiacal signs, at the time 'the child was born' (only rarely is the child named). Nearby dates of the synodic phenomena of the planets are also recorded, along with eclipses, dates of solstices and equinoxes, and lunar visibility phenomena. Occasionally the Horoscope will include a written prediction for the life of the child, such as 'he will find favour wherever he goes', or 'he will have sons', but more often than not no prediction is given. Presumably the interpretation of the astronomical data was communicated orally by the diviner to the customer.

The Development of Astronomical Prediction

Astronomy is not simply an observational science. The advance prediction of astronomical phenomena – eclipses, planetary positions, lunar visibility, etc – plays an equal, if not greater, role in the history of astronomy to that of observation. The motivations for developing methods for making advance predictions are many and varied: to give advance warning of an interesting event to enable preparations to be made to allow easy or accurate observation, and to fill gaps in the astronomical record in the case of bad weather are two of the most obvious reasons. In China, where unexpected celestial events were regarded as bad portents for the emperor, the ability to make accurate advance predictions of an event allowed the heavens to be regulated, negating the importance of the portent.

Three different reasons, along with the general ones listed above, may well have contributed to the drive to develop methods of astronomical prediction in Babylonia. First, advance prediction of an astronomical event had a practical utility: in the case of an eclipse, for example, an advance prediction allowed preparation of the cultic offerings and other practicalities for any required rituals. Second, the ability to calculate astronomical phenomena for any date greatly eased the production of Horoscopes. It is much simpler to calculate, or look up in prepared tables, astronomical data for an individual's date of birth than to search observational accounts, only to discover that the sky was cloudy on the night in question. Finally (but importantly), the development of elegant astronomical theories no doubt arose in part out of intellectual curiosity on the part of the Babylonian astronomers.

Underlying all Babylonian methods of predicting astronomical phenomena is the concept of a period relation. Period relations equate two numbers connected to a phenomenon. For example, the number of years between a certain number of occurrences of a phenomenon before that phenomenon takes place once again at the same location in the sky was one of the most frequently used period relations in Babylonian astronomy. In its simplest form, the period relation was used to predict future dates of first and last visibilities and the other synodic phenomena of the planets in tablets known as Goal-Year Texts.

From their observational record, the Babylonians had noted that after a certain number of years, particular to each planet, synodic phenomena occurred on almost the same date in the Babylonian calendar and at roughly the same location in the sky. These goal-year periods range in length from eight years for Venus to seventy-nine years for Mars. For example, if on the 21st of the second month of a certain year Venus became visible for the first time in the morning, then eight years later on or around the 21st of the second month Venus would again first become visible in the morning. Thus by looking back in the Diaries one period before the 'goal' year for which the scribe wanted to make predictions, he could get a good idea of what was going to happen in the goal year. If this process was performed for each planet, a full set of predictions for the goal year could be obtained. This is just what the Babylonians did in constructing the Goal-Year Texts. These texts are divided into sections for the individual planets, and into each section the scribes copied planetary observations taken from the Astronomical Diary for one period ago. In addition to the synodic phenomena of the planets, passages of the planets by the Normal Stars, lunar and solar eclipses, and lunar six data were included. For Venus, Mercury and Saturn, the observations of the planets by Normal Stars were taken from the same years as the synodic phenomena, but for Jupiter and Mars different, slightly more accurate periods were used.

From the Goal-Year Texts the Babylonian scribes compiled two related types of text known as 'Normal Star Almanacs' and 'Almanacs', applying small corrections to the goal-year data in an attempt to get better predictions. The two types of text share several common features. They both contain predicted data for a coming year, arranged chronologically in monthly sections. Both detail month lengths, the predicted dates of the synodic phenomena of the planets giving the zodiacal sign in which the planet was located at that time, the dates, times and likely visibilities of solar and lunar eclipses, and schematically calculated dates of the equinoxes, solstices and first visibility, acronychal rising, and last visibility of the star Sirius. In addition, the Normal Star Almanacs give predicted passages of the planets by the Normal Stars. This information is replaced in the Almanacs by the dates when the planets entered each zodiacal sign.

Here is an excerpt from an Almanac for year 201 of the Seleucid Era (111–110 BC):

> Month IX, the 1st (of which was followed the 30th of the preceding month). Jupiter in Cancer, Venus in Sagittarius, Saturn in Virgo, Mars in Aquarius. The 3rd, Venus's last visibility in the evening in Sagittarius. The 6th, Venus's [first appearance] in the morning [in ...]. The 11th, Jupiter's acronychal rising. The 12th, Jupiter reaches Gemini. The 13th, Mercury's first visibility in the evening in Capricorn. The 13th, first sunrise before moonset. The 20th, Jupiter [...]. The 25th, Mercury reaches Aquarius. The 2[5th], winter solstice. The 27th, last visibility of the moon. The 29th, Mars reaches Sagittarius.

Although containing related data – indeed, it is likely that the Almanacs were prepared from the Normal Star Almanacs – they may have served different purposes. The Normal Star Almanacs could have acted as guides for observation, telling the astronomers when to look for particular astronomical phenomena and providing surrogate data to be recorded in the Diaries when bad weather prevented observation. Their role, therefore, was much like the 'what to see in the night sky' sections of today's newspapers. The Almanacs, however, provided most of the information necessary for preparing Horoscopes: the zodiacal sign within which a planet lies on a certain day, nearby planetary phenomena and eclipses, etc. They are more akin to the sets of tables used by astrologers to look up where the planets were located on the date of their client's birth.

Prediction of Eclipses

The first astronomical phenomena to be predicted using more refined procedures than simple periodic repetition were, perhaps inevitably given their astrological importance and astronomical complexity, eclipses. As mentioned in the last chapter, during the Neo-Assyrian period some of the scholars who advised the king on the interpretation of celestial omens understood that eclipses of the sun or moon sometimes occur six months after a previous eclipse of the same kind. It is possible that they also knew that eclipses often recur after a period of eighteen years with similar characteristics: about the same proportion of the eclipsed body being covered and the eclipses lasting for more or less the same length of time. However, the second eclipse occurs at about eight hours later in the day than the first. Whether or not the Neo-Assyrian scholars knew this eighteen-year period, known today as the 'Saros', it was certainly used by the astronomers in Babylon by the end of the seventh century BC and probably earlier.

The eighteen-year Saros period, called simply 18 MU.MEŠ '18 years' by the Babylonians, contains 223 lunar months. Combining this knowledge with the awareness that eclipse possibilities usually occur after six months, but occasionally after five months, the Babylonians were able to identify all eclipse possibilities for a long period of time. At first sight this may appear a formidable task, but the method invented by the Babylonians was in fact very straightforward.

First of all, the Babylonians knew that if within 223 months there were a eclipses 6 months after the previous eclipse, and b eclipses after 5-month intervals, then from simple mathematics ($6a + 5b = 223$), a must be 33 and b must be 5. In other words, over a period of 223 months there are 33 eclipse possibilities at 6-month intervals and 5 eclipse possibilities at 5-month intervals. This makes a total of 38 eclipse possibilities within one Saros of 223 months. Second, they knew that eclipse possibilities repeat after one Saros of 223 months. It was therefore simply a matter of distributing the 38 eclipse possibilities as evenly as possible within the Saros, aligning this distribution with records of observed eclipses, and finally repeating this distribution for perpetuity to produce a matrix containing the dates of all future eclipse possibilities. Several fragments of Babylonian texts arranged in matrix-like tables are preserved containing exactly this information.

The Babylonian procedure can be understood in terms of period relations, as the Babylonians certainly knew. The fundamental relationship is:

223 synodic months = 38 eclipse possibilities.

The average interval between eclipse possibilities, the 'period' of the function, is therefore 223 divided by 38, or just over 5;52,6,18 months per eclipse. But since eclipses can only take place when the moon and sun are at syzygy it is necessary to take combinations of some intervals a little over the average and some under the average – at intervals of six and five months. This concept of using intervals above and below the average, preserving the long-term mean, would become one of the main features of Babylonian theoretical astronomy.

The Babylonian methods of predicting eclipses using the matrix-like tables proved both highly effective and very easy to use. A scheme for predicting lunar eclipse possibilities was implemented by at least the beginning of the sixth century BC, and was optimally aligned with the observational record so that it would continue to work down to the end of the fourth century BC before failing to predict an eclipse that could be seen. After that time only small realignments of the matrix were necessary to enable the scheme to keep on functioning. These adjustments were necessary simply because the eighteen-year Saros period is slightly inaccurate, not because of any failing in the principle of the procedure. Perpetuating the scheme to allow future eclipses to be predicted relied simply on the ability to count periods of five and six months, something well within the capability of any scribe or bureaucrat.

Numerical Techniques for Predicting Planetary and Lunar Phenomena

Beginning probably in the fifth century BC, the Babylonian astronomers further developed the concept of period relations to produce advanced numerical methods for calculating planetary and lunar phenomena. We know about these methods from two types of text called 'Ephemerides' and 'Procedure Texts'. The Ephemerides are tables of phenomena calculated by the methods described in the Procedure Texts. However, neither type of text contains an explanation of why the methods work or how they were formulated.

The aims of the lunar and planetary schemes differed. For the planets, the principal goal was to calculate the dates and positions within the zodiac of their synodic phenomena. The movement of the planet in between these phenomena was treated only as a secondary problem. For the moon, the Babylonians' ambitions were much higher. The lunar schemes result in the determination of the length of each lunar month, the date, time and location in the zodiac of the moon at syzygy, the magnitude of any possible eclipse and the duration of visibility of the moon. As we shall see, solving these problems related to the moon required enormous ingenuity.

At the heart of all the schemes for calculating lunar and planetary phenomena lies the same basic principle of the period relation, plus one of two ways of dealing with the variability in the motions of the heavenly bodies. These two methods are called 'step functions' and 'linear zigzag functions', and the planetary and lunar schemes which use the two methods to calculate positions are called respectively 'System A' and 'System B' schemes.

To illustrate the principle of step and zigzag functions, let us look at the case of the monthly movement of the sun through the zodiac. Over the course of one year, the sun appears to pass along a full circle of 360° through the zodiac. The Babylonians knew that there were 12;22,8 months in a year and therefore the sun moved a little more than 29;6,19° per month. However, they also knew that the sun travelled faster in the winter than in the summer. Thus the sun's motion was more than 29;6,19° per month in winter and less than that value in the summer. One approach, therefore, is to assume that the sun moves at a fast speed in one part of the zodiac and at a slow speed in the other part. The monthly progress in the sun's position is determined purely from its initial position. If the sun is in one stretch of the zodiac its monthly progress has one value, if it is in the remainder of the zodiac the progress has a second value. However, if by adding the monthly progress to the sun's initial position the sun crosses from one stretch of the zodiac to the other, then the portion beyond the boundary is modified by a factor equal to the ratio between the two solar velocities. This is what we call a 'step function' (see figure 5). The zones of a step function can be arranged in such a way as to accurately reflect the changing motion of a heavenly body.

A second approach to the problem of the sun's variable motion is to use a function that continually progresses up and down, from a maximum at the point in the year when the sun is assumed to move at its fastest, to a minimum when it moves most slowly. In the case of the sun, the monthly solar progress increases each month by a fixed amount until reaching the maximum where it is reflected about that maximum and decreases every month by the same fixed amount until it reaches a minimum, whereupon it is reflected once more and the whole process starts all over again. This is known as a zigzag function.

Step and zigzag functions are used throughout the lunar and planetary schemes to determine dates and positions in the zodiac, plus variables such as the moon's velocity and latitude. Step functions are, perhaps surprisingly given their apparent simplicity, extremely flexible. Merely by increasing the number and length of the steps it is possible to produce a function that varies considerably.

It is worth noting that the mathematical schemes for calculating planetary and lunar phenomena are purely numerical: they do not rely upon any geometrical or physical model of the heavens or of planetary orbits. Their origin in the mathematics of period relationships distinguishes the Babylonian schemes from later astronomical theories formulated in the Greek-speaking world and in the Middle East during medieval times.

Planetary Schemes

The principal aim of Babylonian planetary schemes was to calculate the date and longitude of all future (or occasionally past) occurrences of the synodic phenomena of the planets given an initial date-longitude pair. The motion of the planets between these phenomena and their variation in latitude was considered only as a secondary problem. The astronomy had to be able to calculate the difference in longitude between two successive occurrences of a particular phenomenon (known as the 'synodic arc'), and the corresponding time interval (known as the 'synodic time'). To simplify this task the Babylonians made the assumption – a fairly good one – that the synodic arc is directly related to the synodic time, so that the latter could be determined just by adding a constant to the former. Thus the planetary systems only had to model the variation in the length of the synodic arc. Several schemes were devised to do this modelling, all of which relied on either the step or the zigzag functions described in the previous section.

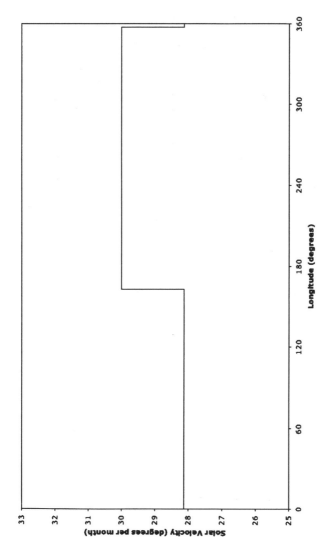

5. A step-function for the velocity of the sun

For the inner planets only step functions seem to have been used (or at least we have no evidence among the preserved texts for use of a zigzag function to determine the synodic arc): they are all of 'System A' type. For each of the outer planets, however, we know of different schemes that use step functions and some that use zigzag functions: both System A and System B models. For some planets several different models are known. For example, we know of five schemes for Jupiter (three of which are of System A variety, the other two being of System B), all of which were being used at around the same date. It would seem that the Babylonian astronomers did not necessarily consider any one of these schemes better than the others, and so worthy of replacement. In fact, from what we know of the Babylonian attitude to astronomy, it seems unlikely that they were interested purely in obtaining the best possible model as compared with their observations. Instead it seems that they thought of all of the models as being equally valid approximations to the way the heavens worked.

The planetary ephemerides are usually arranged with pairs of columns giving the date and longitude of the planet for a series of synodic phenomena. As an example, here are the first few lines from the tablet BM 34621, a System A ephemeris for Jupiter, now in the British Museum. Some of the data on the tablet is broken away, but it is possible to restore the damaged sections with certainty.

2,10 Month XI, 15	1;25 Cancer Second Station	2,11 Month III, 25	14;40 Cancer Last Visibility
2,11 Month XII, 27	1;25 Leo Second Station	2,12 Month V, 7	14;40 Leo Last Visibility
2,13 Month I, 10	1;25 Virgo Second Station	2,13 Month V, 19	14;40 Virgo Last Visibility

2,14 Month II, 22	1;25 Libra Second Station	2,14 Month VII, 1	14;40 Libra Last Visibility
2,15 Month III, 4	1;25 Scorpio Second Station	2,15 Month VII, 13	14;40 Scorpio Last Visibility
2,16 Month IV, 16	1;42 Sagittarius Second Station	2,16 Month VIII, 28	17;36 Sagittarius Last Visibility
2,17 Month VI, 4	7;42 Capricorn Second Station	2,17 Month X, 16	23;36 Capricorn Last Visibility
2,18 Month VI, 22	13;42 Aquarius Second Station	2,18 Month XI, 4	29;36 Aquarius Last Visibility

The first two columns of the table give the dates and longitudes of successive second stations of Jupiter. The second two columns give the same information for Jupiter's last visibility. The date columns give the year number, the month and the day of the month. For simplicity of calculation the months were all taken to be 30 days long, even though about half of all lunar months only last 29 days. However, the error in this approximation is only a fraction of a day and so could be ignored without any loss of accuracy. The longitude columns give the position of Jupiter in degrees within the signs of the zodiac. All numbers are written using the sexagesimal number system.

There exist a few procedure texts that explain how the data on this tablet can be calculated. When fully restored one text begins:

> From 25 Gemini to 30 Scorpio, add 30. From 30 Scorpio to 25 Gemini, add 36. What goes beyond 25 Gemini multiply by 0;50 and add to 25 Gemini. What goes beyond 30 Scorpio multiply by 1;12 and add to 30 Scorpio.

This passage describes a two-zone step function: when Jupiter is between 25° in Gemini and 30° in Scorpio, the synodic arc is 30°, and when it is between 30° in Scorpio and 25° in Gemini, the synodic arc is 36°. We can use this procedure to work through the first longitude column in the table above.

We begin with Jupiter at 1;25° in Cancer. Since this is between 25 in Gemini and 30 in Scorpio, we add 30° to give a new position of 1;25° in Leo. We can continue adding 30° until we reach 1;25° in Scorpio, since each time Jupiter is between 25° in Gemini and 30° in Scorpio. However, adding 30 to 1;25° in Scorpio will take Jupiter past the zone boundary at 30° in Scorpio. According to the rule in the procedure text, we must therefore multiply the 1;25° that would exceed the zone boundary by 1;12° to give 1;42°. Adding 1;42° to 30° in Scorpio, we arrive at 1;42° in Sagittarius, which is the figure given in the text. Jupiter is now between 30° in Scorpio and 25° in Gemini, so the procedure text tells us that we must add 36° each line. The next entry is therefore 7;42° in Capricorn. Exactly the same procedure may be followed to produce the longitudes of last visibility given in the final column of the table.

To obtain the corresponding dates of the phenomena, another procedure text says

> From (one) appearance to (the next) appearance, put down (the arc) between them. You add 12;5,8,8,20 to it and predict the date.

This particular passage describes what to do to calculate the dates of the first appearance of Jupiter, but the same procedure is used for all of the phenomena. All you do to obtain the synodic time is simply to add 12;5,8,8,20 to the synodic arc, plus an additional 12 months since the sun must also make one complete revolution of the Earth. We can try this rule out on the dates in the table above.

We begin with Jupiter's second station on the 15th of Month XI in year 2,10. The synodic arc between this and the next second station is 30°. So we add 30 to 12;5,8,8,20 to obtain 42;5,8,8,20. The synodic arc is therefore 12 months plus 42;5,8,8,20 days (remembering that we always assume 30 days in a month during the calculation), which is equal to 13 months plus 12;5,8,8,20 days. Adding 13 months to 15th of Month XI in year 2,10 brings us to the 15 of Month XII in year 2,11, and then adding 12;5,8,8,20 days we come to 27;5,8,8,20 days in Month XII, which we simply round down to obtain the 27th of Month XII in year 2,11. The same procedure can be followed to obtain all of the dates in the text.

Lunar Schemes

The Babylonian schemes for calculating lunar phenomena were more ambitious, and show much greater ingenuity in their construction than their planetary models. These schemes allowed the Babylonian astronomers to calculate the interval between one conjunction of the sun and moon and the next. This period is called the 'synodic month' and varies from about 29¼ to a little over 29¾ days. When combined with methods of working out the conditions under which the new moon will be visible, the variable length of the synodic month could be used to generate the calendar. In addition, the schemes predicted in which months eclipses of the moon and sun would take place and at what time of the day or night.

The two main schemes for calculating lunar phenomena, System A and System B, share a common basic structure and approach but display many differences in how they solve the same problems and in the period relations and parameters which underlie the schemes. Both System A and System B ephemerides contain many columns, more than fifteen in some cases, that are needed to produce the final output of the ephemeris: the length of the month, the visibility of the moon around new and full moon, and the circumstances of possible eclipses. A photograph of a fragmentary System A ephemeris is shown in figure 6.

One of the remarkable features of the lunar schemes is the development of methods of coping with highly irregularly variable properties of phenomena by splitting them into two or more simple functions and adding them together. For example, the length of the synodic month depends upon the speeds of the sun and moon, which vary over different time periods: the cycle of the sun's speed repeats after one year of 365 and a fraction days, the moon's after about 27½ days. Each variation can be approximated fairly well using either a step or a zigzag function, and their relative contributions to the variation in the length of the month can then by obtained by appropriate scaling and adding.

The lunar theories, in particular System A, are far more complex and subtle than can be described here: consult some of the works cited in the bibliography for a full discussion. The development of the two lunar schemes reveals a truly remarkable application of mathematics to astronomy. They, or at least System B, were known to the later Greek astronomers such as Hipparchus and Ptolemy, who based important parts of their own astronomy on these Babylonian methods.

Uses of Babylonian Numerical Astronomy

One curious fact concerning Babylonian numerical astronomy continues to puzzle historians: we have almost no evidence to show whether any of it was used for anything. Many ephemerides containing calculated phenomena are known, but this calculated data is not the source of the predictions found in the Diaries and Almanacs, which apparently originate from the Goal-Year Texts. Why go to all the trouble of formulating these extremely complicated numerical

6. Fragments of a Babylonian ephemeris for the moon

methods if they were not to be used? This question exposes a fundamental gap in our understanding of Late Babylonian astronomy.

One possibility for why the astronomers tended to use the goal-year techniques rather than the numerical schemes may have been simply that the former are much easier to use and yield results that are not significantly poorer in accuracy. This is not to say that the numerical schemes were bad, but rather that the goal-year methods work surprisingly well. The calculation of planetary first and last visibilities, in particular, are probably predicted to an accuracy of within ten days or so by both methods, often much better. This level of accuracy may not appear to be very good at first sight but, even today, using the latest astronomical models, it is only possible to predict the dates of planetary visibilities within a range of a few days. The uncertainty is caused by variable and unpredictable local conditions at the horizon: changes in temperature and humidity affect atmospheric refraction (the bending of light near the horizon) and extinction (the dimming of light travelling through the atmosphere); the acuity of the observer's eye means that one person may be able to see a dim planet whereas another may not; and clouds or bad weather may prevent anything at all being seen in the sky.

A recent discovery has shed new light on the issue of whether the numerical planetary and lunar schemes were ever used. A unique tablet in the collection of the Oriental Institute in Chicago, where it has been given the accession number A 3405, contains a collection of planetary and lunar phenomena calculated using the numerical schemes and arranged chronologically for an eleven-year period. The tablet was originally owned by a well-known individual from the city of Uruk in southern Mesopotamia named Anu-bēl-šunu, and was written by his son Anu-aba-utēr. Both father and son owned or wrote a number of astronomical and astrological texts as well as appearing as witnesses in several business transactions. Anu-bēl-šunu's text is particularly interesting because the data it contains was calculated for the years 60 to 70 of the Seleucid Era (252–241 BC). However, the text was written in year 121 of the Seleucid Era (191 BC). Why did Anu-bēl-šunu calculate planetary and lunar phenomena for more than fifty years earlier? It may have been to provide data for constructing horoscopes. By a remarkable chance, Anu-bēl-šunu's horoscope survives and places his date of birth in year 63 of the Seleucid Era (249 BC). Did Anu-bēl-šunu make this unique text to allow himself to cast his own horoscope? Unfortunately, we cannot answer this question with any surety, but a link between the numerical astronomical schemes and the practice of horoscope casting seems possible. As described in the next chapter, casting horoscopes was the principal use of astronomical methods in the Greco-Roman world, where there was considerable Babylonian influence.

Astronomy in the Greek and Roman Middle East

In 331 BC Alexander the Great marched into Babylon and brought Mesopotamia under Greek rule for the first time. The Babylonian Astronomical Diary for that year records the event as follows (sadly the text is fragmentary):

> (Month VI) That month, on the 11th, panic occurred in the camp before the king. ... On the 24th, in the morning, the king of the world (Alexander) [...] they fought with each other, and a heavy defeat of the troops of [...] the troops of the king deserted him and [went] to their cities ... (Month VII) On the 11th, in Sippar an order of Al[exander ...] (saying) "I shall not enter your houses". ... Alexander, king of the world [came in]to Babylon.

In the following years Alexander returned frequently to the city in between marching further and further eastwards into central Asia. Eventually he would die there, the Diary for 323 BC reporting rather prosaically 'the 29th, the king died, clouds'. Bringing Mesopotamia into the Greek empire enabled Greek-speaking astronomers to interact with the Mesopotamian tradition of astronomy.

Knowledge of observations made by Babylonian astronomers, along with their methods of predicting astronomical phenomena, was very important to Greek astronomers. There was no tradition of making or recording regular astronomical observations in Greece, so that later Greek astronomers had very few observations made by their predecessors to work with. By combining what they knew of Babylonian astronomy with the early Greek tradition of philosophical speculation into the nature of the universe, later Greek astronomers were able to break new ground in the development of theories for modelling the motions of the planets. Two astronomers in particular, Hipparchus and Claudius Ptolemy, used Babylonian empirical data and elements of Babylonian astronomical theories alongside the few available observations and ideas of earlier Greek astronomers to produce a new system of astronomy that treated the motion of the planets as a geometrical problem. Ptolemy's book on astronomy, known today as the *Almagest*, provided the bedrock on which almost all later astronomers founded their studies until Johannes Kepler proposed a new approach to astronomy in the seventeenth century AD. Later Greek writers, Islamic astronomers and the main figures in the European Renaissance such as Regiomontanus and Copernicus all did their work in refining models for the motions of the sun, moon and planets in the context of Ptolemy's book.

That is not to say that everyone accepted Ptolemy's models uncritically. As discussed in a later chapter, some medieval Islamic astronomers were very critical of parts of Ptolemy's work. Even in his own time and for the next couple of centuries, Greek astronomers identified problems in Ptolemy's astronomy and, as has only recently been discovered, many practising astrologers in Roman Egypt during the second to fourth centuries AD used Babylonian planetary schemes to calculate astronomical data for horoscopes.

A Philosophical Approach

Before the fourth century BC Greek interest in astronomy was primarily concerned with questions of cosmology. How was the universe created? Is the universe infinite or finite in size? What is the Earth's position in the universe? What shape does the Earth have? In attempting to answer these questions, early Greek writers largely eschewed the idea of making detailed observations of the heavens and relied instead on what was widely known from everyday experience and their sense of reason. Early Greek astronomy was basically the preserve of philosophers thinking during the daytime, not astronomers looking up to the sky at night.

All accounts of the views of Greek philosophers from before the fourth century BC about the structure of the universe are lost in their original writings. We only know of the ideas of Thales or Pythagoras, for example, because they were transmitted to us by later writers. This can lead to problems of misrepresentation. Thales of Miletus, for example, was credited by the Greek historian Herodotus with predicting a change from 'daylight to darkness' during a battle between the Lydians and Medes in around 585 BC. Herodotus was writing a century after the event and his words are very vague. By the time later Greek writers describe the event, the 'darkness' had been transformed into an eclipse of the sun. It is certain, however, that Thales had no means with which to predict the path of a solar eclipse, and that whatever he did do during the battle was exaggerated by later authors. It is evident from the stories about Thales that he was built up to be an astronomical genius – after all, according to Plato he was reported by a 'pretty and clever maidservant from Thrace' to have been so busy gazing at the stars when out walking that he fell down a well!

Even more problems exist with references to Pythagoras. Today, his name is familiar to schoolchildren throughout the world because of its association with the mathematical rule that equates the square of the hypotenuse of a right angle triangle with the sum of the squares of its remaining two sides. In the ancient world, however, his name also evoked the image of a white-clothed mystic with a mysterious golden thigh, a man who could converse with beasts and convince a wild bear not to eat other animals, a trainer of athletes, a philosopher who proposed that the universe was founded on a cosmic harmony, and the founder of a secret order who exercised considerable political power, to cite only a handful of the claims made about him by later authors. Distinguishing between truth and fiction in these accounts is almost impossible. Regarding his astronomy, all we can extract from the writings about him is that he believed that there existed a numerical beauty to the universe, just as numerical relationships underlie musical intervals that are pleasant on the ear, and that he or his followers believed that the Earth, sun and planets move in circles around a central fire.

The culmination of the Greek philosophical approach to astronomy came with the writings of Aristotle. His views became almost universally accepted by later Greek thinkers and subsequently in Europe and the Middle East. Aristotle argued that the Earth formed a stationary sphere at the centre of the universe. In support of this idea he gave examples from everyday experience. For example, the Earth must be at rest because weights thrown upwards come back down to the same place. It must be spherical because the Earth's shadow as seen on the moon during an eclipse is always curved, and because as we move northwards or southwards we do not all see the same stars.

In Aristotle's universe there is a strict separation between the things that lie beneath the moon and those that lie above it. In the sublunar region the Earth and its inhabitants (both thinking and inanimate) are made up of combinations of four elements, earth, air, fire and water. They are naturally at rest or moving in straight lines. The heavens are made up of a fifth element, quite different from those found on the Earth. This fifth element, the *quintessence*, has a natural circular motion. Therefore, every movement in the heavens is made up from circular paths. This key belief would drive all subsequent attempts to deduce mathematical theories of the motion of the sun, moon and planets – and cause no small problem for the astronomers who developed these theories.

Although Aristotle's view of a universe with the Earth stationary in the centre was almost universally accepted, one interesting alternative model deserves mentioning. Aristarchus of Samos claimed that the Earth was not stationary and at the centre of the universe. Instead, the Earth rotated on its axis daily and, more strikingly, it moved around the sun, which held the central place in the universe. In making this claim, Aristarchus put forward the very same model that Copernicus proposed in the sixteenth century AD. Also as in the case of Copernicus's theory, his idea met with a hostile reaction in some quarters. According to Plutarch, a stoic philosopher named Cleanthes wanted Aristarchus to be charged with impiety for placing the Earth in motion. Unlike Copernicus, however, who found champions in Galileo and Kepler, Aristarchus's idea found almost no supporters. His own writings on the subject are lost and the few other writers who even mention Aristarchus's theory uniformly reject it.

Geometry in the Heavens

The challenge facing later Greek astronomers was to reconcile

Aristotle's model of the universe with observations to try to formulate mathematical theories of the motions of the heavenly bodies. These theories would allow the positions of the sun, moon and planets to be calculated at any given time. The Aristotelian approach required that Earth be placed stationary at the centre of the universe and all motions in the sky be explained by means of circles. The first restriction, a central, stationary Earth, was easily accommodated: the movement of the sun and the stars across the sky every day was simply attributed to a uniform daily motion of everything in the sky. The problem arose with the second part of Aristotle's stricture: it was well known that the planets did not move uniformly with respect to the fixed background of stars. For example, during some period of the year Mars, Jupiter or Saturn may change direction and move backwards in a loop, as Aristotle himself was well aware. How could this be accounted for using only circular motions?

Several astronomers proposed combining two or more circular motions of different sizes and speeds for each planet to produce an overall, apparent motion that was not circular. For example, Eudoxus, whose model Aristotle described, suggested that planetary orbits could be explained by assuming that the planet's path was traced by a point on the equator of one sphere rotating with uniform velocity, this sphere itself rotating with the same velocity but in the opposite direction and tilted to one side. The path traced by such a model resembles a figure-of-eight – not a very good approximation to a planetary orbit, but better than a simple circle.

Real progress in developing models for the movements of the sun, moon and planets came about through the work of two of the most important Greek astronomers, Hipparchus of Rhodes and Claudius Ptolemy. Hipparchus, who was born in Nicaea in Asia Minor in the early part of the second century BC, lived for most of his life on the island of Rhodes just off the coast of modern Turkey. According to local legend, he made his observations of the sky from the top of the hill above the east of the old city, although we have no evidence for this in his own writings. Unfortunately, all but one of his books are lost. The only surviving work is a commentary on a poem by Aratus that describes the grouping of stars into constellations. The book tells us less about his astronomy than about his attitude to writers he considered to be incompetent: in tone it is highly critical, in parts intolerantly so. Fortunately, Claudius Ptolemy describes several aspects of Hipparchus's astronomy in his book known as the *Almagest*.

Ptolemy lived in Alexandria in Roman Egypt during the second century AD. He wrote many books, most of which are preserved, on topics including mathematical geography, astrology, optics, musical harmonics and astronomy. His major astronomical work is a thirteen-book treatise on mathematical astronomy, known today by a Latinized version of its Arabic title the *Almagest*. In it Ptolemy sets out a detailed and orderly treatment of his mathematical models for the movements of the sun, moon and planets around the Earth. He begins by describing and justifying the basic principles of the Aristotelian universe, and then outlines the mathematical methods he will use throughout the book. Next he describes how to use observations to derive his models, before finally presenting the models themselves and explaining how they can be used to predict the positions of the planets. Ptolemy often refers to the work of Hipparchus, and builds upon his work. Interestingly, he mentions no other astronomers who tried to develop mathematical models by name: it seems he deemed only Hipparchus to be worthy of that honour.

The main tool used by both Hipparchus and Ptolemy in their astronomy was the geometrical hypotheses of epicycles and eccentric circles. The origin of these ideas is not known, but it must have been realized well before Hipparchus's time that the two hypotheses could account for the variable velocities of the sun, moon and planets and, if arranged correctly, the retrograde motion of the planets. To illustrate the epicycle and eccentric models, here is a simple example of the way in which they were used to model the variable motion of the sun.

The sun appears to move around the Earth along a great circle called the ecliptic. Because the ecliptic is tilted at an angle of about 23° to the celestial equator during the summer, the sun will be higher in the sky and daylight will last longer than in the winter, when the sun will be lower in the sky and daylight will be shorter. The dates of longest and shortest daylight are called the summer and winter solstices and the two days during which daylight is the same length as night are called the equinoxes. The position of the sun in longitude on the days of the solstices and equinoxes divides the ecliptic into four equal parts. If the sun's motion along the ecliptic is assumed to be a uniform circle centred on the Earth, then the division of the ecliptic into equal parts will be as shown in figure 7a. The seasons (defined as the number of days between a solstice and an equinox or vice versa) will simply be one-quarter of the length of the year, or roughly 91⅓ days long. However, by carefully observing the dates and times of the solstices and equinoxes, Hipparchus found that spring lasted for 94½ days, summer for 92½ days, autumn 88⅛ days, and winter 90⅛ days. How could one account for the sun moving faster along some parts of the circle and slower along others?

One way around the problem is to move the Earth slightly away from the centre of the circle (see figure 7b). The four equinox and solstice points are still at right angles from each other as seen from the Earth, but now the sun must travel further around the circle between the points marking the vernal equinox and the summer solstice. Hipparchus adopted this model – called an 'eccentric' model because the centre of the sun's path is displaced slightly from the central Earth – and used observations to determine four

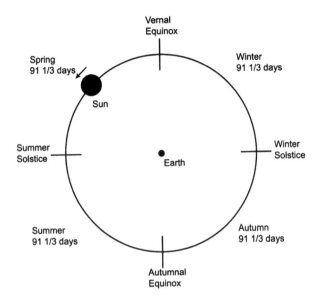

7a. The development of Hipparchus's eccentric model for the motion of the sun. The Earth is at the centre of the circular path of the sun, so the length of the four seasons is equal

parameters that governed the system: the length of the tropical year which is equal to the period of the sun's motion on the circle, the direction of the line from the centre of the Earth to the centre of the circle upon which the sun moves, the relative distance of the Earth from the centre of the circle compared with the radius of the circle, and an initial longitude from which to start any calculations.

Although Hipparchus used the eccentric model for the sun's motion, he knew that there was an equivalent alternative: the epicyclical model (see figure 8). In this model the Earth returns to the centre of one circle, called the 'deferent', but the sun moves on another smaller circle, called the 'epicycle', whose centre is fixed to the deferent circle. In Hipparchus's solar model, the two circles rotate with the same speed but in opposite directions. It is easy to see that under this restriction the eccentric and epicycle models are geometrically identical.

Hipparchus used similar models for the moon. However, whereas the various parameters for the model of the sun could easily be obtained from observations of the times of solstices and equinoxes, for the moon there is no similar simple set of observations that can be made. Instead, Hipparchus drew on his knowledge of Babylonian astronomy for some of the required parameters, and devised ingenious means of using Babylonian and Greek observations of

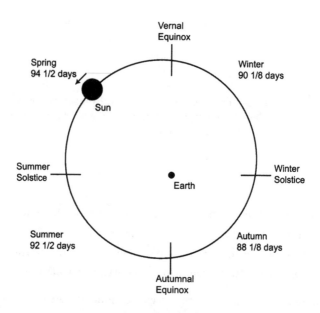

7b. The Earth is moved off-centre so that the four seasons are the correct length

lunar eclipses to obtain the others. How Hipparchus came by his knowledge of Babylonian astronomy is not known. Did he visit

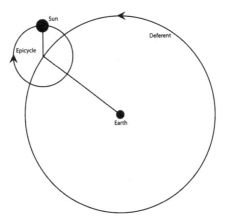

8. An epicycle model for the sun

Babylon himself and speak to the astronomical scribes there? Or did one of these scribes find his way into the Greek world, bringing texts containing astronomical records with him? We shall probably never know the details, but the translation of Babylonian texts into Greek must have involved the collaboration of Babylonian astronomers, for only they would have understood the unusual terminology that the texts contain.

Despite his many innovations in establishing parameters, Hipparchus's lunar theory was in the end not particularly successful. His basic premise that the moon's motion could be modelled using a simple epicycle or eccentric theory was wrong. Observations showed that although his model worked well enough when the moon was in conjunction or opposition with the sun, at other times during the moon's orbit its calculated longitude was often considerably off. Ptolemy introduced a revision to Hipparchus's lunar theory in which the moon's epicycle is carried on a circle whose own centre moves around the Earth in the opposite direction (see figure 9). Although Ptolemy's revision had the desired effect in producing positions of the moon which agreed well with observation, it suffered from one big problem, a problem that Ptolemy himself is strangely quiet about: in Ptolemy's model the distance of the moon from the Earth varies by a factor of about two, or in other words, the size of the lunar disc will at times be double its size at other times. Anyone who looks at the moon every night will soon realize that this is nonsense – the moon barely appears to change size at all. A solution to this problem was not found until the thirteenth century, when the Islamic astronomer Nasīr al-Dīn al-Tūsī invented the 'Tūsī couple'.

For the planets, Ptolemy found that the simple eccentric or epicycle model was also insufficient to account for their variable motion. Instead, he used a combination of the epicycle and the eccentric model in which a planet moved on an epicycle to a deferent circle

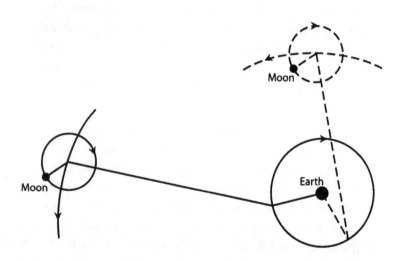

9. Ptolemy's final lunar model. The moon moves on an epicycle, whose centre moves on a deferent, whose own centre moves around the Earth on a circular path. The 'crank' caused by the inner circle means that the moon varies considerably in distance from the Earth

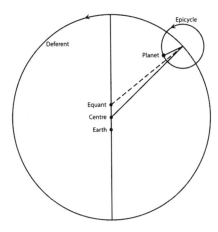

10. Ptolemy's equant model for an outer planet. The planet moves on an epicycle whose centre moves around a deferent in such a way that as viewed from the 'equant point' it has constant speed. The 'equant point' is opposite the Earth from the centre of the deferent circle

whose centre was eccentric to the Earth. He also added a further complication – a mathematical device that has become known as the 'equant' – and in doing so broke one of the fundamental rules of the Aristotelian universe: uniform circular motion. The equant is a point located away from the centre of a circle about which a point on the circle moves with constant angular velocity. Because this motion is constant as seen from this point, it cannot be constant from the centre of the circle. Ptolemy placed his equant point opposite the Earth from the centre of the circle (see figure 10).

In using the equant in his models, Ptolemy left an ugly chink in the physical and philosophical rationale of his astronomy. Although it served its purpose perfectly in providing a significantly improved method of calculation, the equant could not be explained physically. If the circles along which the planets moved were real objects – it was often claimed that they were solid crystalline spheres – how could they rotate with uniform angular velocity about a point that is not its centre? The answer is simple – they cannot. Philosophers and astronomers would gradually pick away at this apparent flaw in Ptolemy's astronomy for the next thousand years and more until a way was finally found to replace the equant with a combination of circular motions by Islamic astronomers in the thirteenth century.

Astronomy in Greek and Roman Society

Within the pages of Ptolemy's *Almagest* there is little or no indication of what he intended his astronomy to be used for. However, in another work Ptolemy says that there are two means of prediction in astronomy. First, and most certain, is the ability to calculate the positions of the sun, moon, planets and stars, the topic dealt with, he says, in its 'own treatise', the *Almagest*. The second aspect of astronomical prediction, however, is the effect of the positions of the heavenly bodies on the world – in other words, astrology. It is astrology that is the subject of the treatise he was now writing, a work in four books that is now known as the *Tetrabiblos*. There is no doubt that astrology was the primary use of astronomy in Greek and Roman society. While the philosophers may argue over the nature of the universe, and the mathematicians over how best to model planetary motions, for most people what mattered was the ability of astrologers to cast and interpret horoscopes.

A horoscope consists of a list of the longitudes of the five planets, the moon, the sun and the point on the ecliptic (called the '*horoscopus*') that is rising at the moment of a child's birth. A few hundred Greek horoscopes are preserved on papyri and in literary sources. Some flesh out the minimum astronomical information listed above with the name of the child, the date and time of his or her birth, and sometimes an interpretation of the astronomical data to predict the fate of the child.

Very few of the street-corner astrologers who could cast your horoscope for a fee would know, or care, about the details of theoretical models for predicting planetary positions. What the astrologers needed were simple procedures that they could apply to determine the longitudes of the sun, the moon, the planets and the *horoscopus* for the moment of a customer's birth. A long book such as Ptolemy's *Almagest*, which went into laborious detail of how astronomical theories were constructed, was of little use. Instead, astrologers relied upon astronomical tables where one could look up a few entries using the date of birth, add them together, and out would come a planet's longitude. Ptolemy and other astronomers produced aptly named *Handy Tables* of this form, which astrologers could use without having to understand the astronomy on which they were formulated.

A fascinating insight into the everyday practice of astrologers in a provincial town in Roman Egypt has recently been provided by the discovery and analysis by Alexander Jones at the University of Toronto of a large body of astronomical texts written in Greek on papyri. These papyri were among the material excavated from the rubbish dump of the city of Oxyrhynchus in Egypt by two English archaeologists named Grenfell and Hunt between 1896 and 1906. Jones found that the vast bulk of the astronomical papyri are either horoscopes or tables that could be used for casting horoscopes. Only a very small fraction consists of prose texts explaining how to produce astronomical tables or the theory behind them. Even more interestingly, it turns out that many of the astronomical tables are not computed using Ptolemy's astronomical theories or anything similar, but use Babylonian astronomical methods. The papyri provide a clear illustration of the indebtedness of Greek astronomy to the Babylonians. Evidently the astrologers felt no nationalistic bond to Greek astronomy. They simply used whatever methods were available and most convenient to produce horoscopes.

The End of Greek Astronomy

Ptolemy's *Almagest* marked the culminating point of Greek astronomy: following its publication little new work was achieved by Greek astronomers. Several authors wrote commentaries on Ptolemy's work, explaining its concepts and methods, illustrating mathematical procedures and occasionally proposing minor changes, but within a few centuries Greco-Roman civilization was in the midst of a terminal decline and many scientific works were beginning to be forgotten in Europe. However, some of these treatises found a readership in India and the Middle East, where they were copied and sometimes translated into local languages. Some of Ptolemy's works, for example parts of his book on the physics of the heavens, the *Planetary Hypotheses*, were completely lost in Greek and now only survive through medieval Arabic translations.

As described in the remaining chapters, astronomers in the medieval Islamic world took Greek astronomy and combined it with the native traditions of the Middle East to put astronomy to new uses. In the process they introduced many innovations to resolve what were seen as problems in Ptolemy's theories, made new observations to correct various parameters and in turn passed some of these improvements on to the astronomers of the European Renaissance such as Regiomontanus and Copernicus.

Astronomy in Medieval Islamic Society

The founding of the Islamic faith and its rapid spread through the Middle East in the decades following the death of the Prophet Muhammad in 632 AD brought with it a new and important role for astronomy in everyday life. While in Mesopotamia astronomy was useful for helping to regulate the calendar and providing observations of the heavens that could be interpreted for the king, and in the ancient Greek world astronomy played a central role in many philosophical arguments (not to mention keeping numerous astrologers in food and wine), no society has embraced and relied upon astronomy as much as the Islamic nations.

Astronomy plays an essential role in the practice of the Islamic faith in three main ways. First, observation of the new moon crescent determines the beginning of each month, crucially important in particular for the beginning of Ramadan, the month of fasting. Second, the five daily prayers are to be performed at moments during the day that are defined by the shadow cast by a gnomon or the position of the sun overhead. Finally, prayers should be performed facing the sacred Ka'ba in Mecca, a direction known as the *qibla*.

Since the direction to Mecca changes depending upon where one is located upon the Earth's surface – in London it will be southeast, but northwest in Sydney, Australia – determining the *qibla* can be solved by the methods of spherical astronomy.

The importance of astronomy for the correct ordering of religious observances provided an environment within which astronomy and astronomers could flourish. Some found employment in mosques, holding the title of *muwaqqit*, where they were to advise the muezzin of the times of prayer. Others worked in observatories set up by local rulers. Whether they were employed in official capacities or not, many astronomers prefaced their writings with statements pointing to the usefulness of astronomy. Perhaps nowhere else, before or since, have astronomers been considered such useful members of society.

Specialist astronomers did not, however, have a monopoly on observing the heavens. Alongside the technical astronomical tradition, there existed simpler methods for solving the astronomical problems of religious practice. These methods, called 'folk astronomy' by the respected historian of Islamic astronomy David King, did not rely upon complex mathematical models of the heavens. Instead, simple observational techniques that gave good approximations to the *qibla* and the times of prayers were often considered adequate enough. Since technical astronomy was forever associated with its disreputable twin, astrology, and furthermore was founded upon methods developed by infidel ancient Greek scientists, many Muslims preferred this untainted folk astronomy. Islamic legal scholars, for example, seem to have advocated folk astronomy methods. But even if its association with astrology had not been an issue, technical astronomy was simply too difficult for most people to understand. Folk astronomy was therefore often preferred by the general population.

The Islamic Calendar

Before the rise of Islam the calendars used by people in the Middle East were of the same kind as that used in ancient Mesopotamia. The beginning of each month was defined by the sighting of the thin crescent of the new moon, there were twelve months in a year and, in order to keep the year in line with the seasons, extra 'intercalary' months were added once every three years or so. However, in the *Qur'an* the practice of intercalation is expressly forbidden, apparently because the insertion of an extra month in a year could mean that particular months that had been intended to be holy could be confused with other months.

Lunar months, whose beginning is determined by the first sighting of the new moon crescent, can last for either twenty-nine or thirty days. A year of twelve months therefore contains either 354 or 355 days. The difference between 354 or 355 days and the solar year of a little under 365¼ days is about eleven days. This difference means that each year a date in the Islamic lunar calendar will slip back in our Gregorian calendar, which keeps up with the seasons, by these eleven days. For example, the beginning of Ramadan, the month of fasting, takes place earlier and earlier in the solar year as time goes on. Here are the approximate dates of the beginning of Ramadan in 2004–2008:

16 October 2004
5 October 2005
24 September 2006
13 September 2007
22 August 2008

In approximately thirty-three years the beginning of Ramadan will have moved through the whole cycle of the seasons, from the long, hot days of summer, to the short, cool days of winter, when going without food or water during the hours of daylight must be much easier.

The dates of Ramadan given above are only approximate because the beginning of the lunar month is announced by Muslim religious authorities on the basis of observation of the new moon crescent. It might be imagined that nothing could be simpler, as long as the weather is clear. But the visibility of the thin new moon crescent depends upon several factors, some of which are changeable from place to place, such as the atmospheric refraction that bends light near the horizon and varies with temperature and humidity. Other factors are subjective, such as the acuity of the observer's eye. An even bigger problem is false sighting of the moon's crescent.

In the 1990s two American astronomers, Leroy Doggett and Bradley Schaefer, arranged for more than 2,500 amateur astronomers to watch for the new moon crescent. They found that around 15 per cent of the observers honestly thought they had seen the new moon on occasions when it was impossible for them to have done so. The human mind often sees what it expects to see. This problem was known in medieval times as well. The diaries of Muhammad ibn Jabayr describe his journey of pilgrimage from Granada in southern Spain to Mecca and back towards the end of the twelfth century. Ibn Jabayr described the scene where, one day in Mecca, a crowd of observers was watching for the new moon crescent. A cry suddenly went up from one man who had spotted the crescent, followed by further cries from other observers. However, when a group went to the judge to say that it had seen the crescent, the judge could not conceal his disbelief, for that evening the sky was full of clouds. No one could see the sun behind those clouds, he retorted, never mind the thin crescent moon!

It was in order to eliminate some of the problems in deciding whether the new moon had been seen that astronomers could be called upon. In the preface to his *Hākimī zīj* (a *zīj* is a collection of astronomical tables with explanations), the astronomer Ibn Yūnus wrote that astronomy and the observation of the heavenly bodies are associated with religious law, as astronomy allowed the beginning of months to be determined. Many medieval Islamic astronomers tried to develop accurate criteria for the visibility of the new moon crescent. Today the problem is still not fully solved although good models for predicting crescent visibility have been produced, principally by the Malaysian astronomer Mohammad Ilyas.

Al-Khwarizmī, more famous for his work in mathematics – our term 'algorithm' derives from his name – was one of the earliest Islamic astronomers to study the problem of lunar crescent visibility. He understood that the visibility of the moon's crescent depended upon two principal factors: is the moon sufficiently distant from the sun, and is it high enough in the sky to be seen in the background glow of dusk? Adopting an Indian idea that the crescent would be seen if it set at least 48 minutes after the sun, al-Khwarizmī set out to calculate the setting times of the sun and moon. This was no trivial task as the daily rotation of the Earth is not in the same plane as the apparent movement of the sun and moon through the zodiac. In fact, at different times of year, the ecliptic – the path of the sun and roughly the centre of the moon's path – cuts a steeper or shallower angle to the horizon than the celestial equator, the direction in which the heavens rotate. Furthermore, the celestial equator cuts the horizon at different angles depending upon the geographical latitude. To complicate matters even further, the moon moves in latitude perpendicular to the ecliptic as well as progressing through the zodiac (see figure 11). Determining the length of time between sunset and moonset was therefore a complicated problem of spherical trigonometry.

Al-Khwarizmī's solution to the visibility problem was to produce a table that gave for each zodiacal sign the distance between the sun and moon along the ecliptic that corresponded to a difference in setting time of 48 minutes. The table was constructed for use at the geographical latitude of Baghdad, al-Khwārizmī's home town, and must have been quite popular as it is preserved in three manuscripts containing the works of other astronomers. However, it has one major flaw: al-Khwārizmī did not take into account the motion in latitude of the moon. This problem was rectified in later tables, such as that by the tenth-century Egyptian astronomer Ibn Yūnus.

Almost all medieval Islamic astronomers devised tables for determining the visibility of the new moon crescent. Some of these show a remarkable understanding of the phenomenon, incorporating features such as the profile of the local horizon (mountains, valleys, etc) as well as the velocity of the moon and even the variable amount of light in the crescent due to its 'age' since conjunction with the sun.

Their works illustrate not only the importance of the problem of

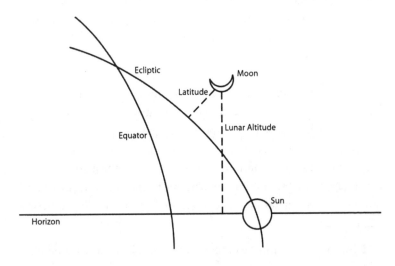

11. The horizon at sunset on the day of first visibility of the lunar crescent

determining the beginning of the month for Islamic society, but also the high level of astronomical understanding among medieval Islamic astronomers.

The Times of Prayer

Five times every day the muezzin of a mosque calls local Muslims to prayer. In doing so he enables Muslims to perform their prayers at appointed times during the day and night, following a tradition that stretches back to at least the eighth century AD. Today the muezzin probably reads the times of the prayers from an approved handbook. The times themselves, however, are defined by the height of the sun in the sky. Because the sun's highest position in the sky changes day by day, and also as one moves north and south on the Earth's surface, the times of the prayers vary during the year and with geographical location. Many Islamic astronomers attempted to solve the problem of determining the times of prayers without observations – useful, for example, on cloudy days when the sun could not be seen. By and large, however, the methods advocated by astronomers were rejected by the legal scholars. For example, Ibn Rahīq, an eleventh-century author living in Mecca who wrote a treatise on the calendar and timekeeping, stated that the times of prayers should only be determined by observation with one's own eye, not with 'any of this astronomy nonsense'.

The standard definitions for the intervals of time during which it is permitted to make the five daily prayers were adopted in the eighth century. They are as follows. The first prayer (*maghrib*) must be performed between sunset (the beginning of the day in the Islamic calendar) and nightfall (the end of dusk). The interval for the second prayer (*'ishā'*) is between nightfall and when either one-third or one-half of the night is past. The third prayer (*fajr*) should be performed between daybreak (the beginning of dawn) and sunrise. The fourth prayer (*zuhr*) begins when the sun has crossed the meridian or when the shadow of an object has been observed to increase (the two are astronomically identical) and must be completed before the beginning of the fifth prayer (*'asr*), which takes place at the moment when the shadow of a gnomon has increased by the length of that gnomon. The allowed time period for the fifth prayer ends either when the shadow increase is twice the length of the gnomon or at sunset. In some parts of the medieval Islamic world *zuhr* begins when the shadow of the gnomon has increased by one-quarter of the length of the gnomon.

The times of *zuhr* and *'asr* are defined by the height of the sun in the sky. At midday the sun is at its highest point in the sky, due south for those of us in the northern hemisphere, and a shadow cast from a gnomon, a stick or rod placed vertically on flat ground, will be at its shortest. As afternoon progresses the shadow cast by the gnomon will gradually increase in length and swing round towards the east (see figure 12). By measuring the increase in length of the shadow the prayer times can be determined.

Recent work by historian of Islamic astronomy David King of the Goethe University in Frankfurt has found that the times defining the two afternoon prayers were originally conceived from the notion of 'seasonal hours'. In contrast to our 'equinoctial' or 'equal' hours, which last one twenty-fourth of the period from one midnight to the next, seasonal hours divide the period of daylight into twelve equal parts and similarly the night into twelve parts. In the winter this means that an hour during the day is shorter than a night hour; in the summer it is the opposite. Only on the dates of the equinoxes will daytime hours be the same length as night-time hours. David King has discovered that the times of *zuhr* and *'asr* correspond to the sixth and ninth hours of daylight according to a simple formula for a gnomon that was widely known in the Islamic world in the eighth century but came originally from India. Seasonal hours were widely

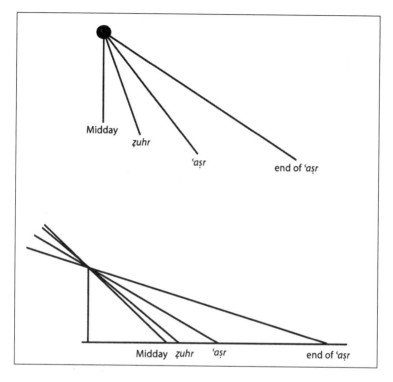

12. The shadow cast by a gnomon as seen from above and the side at the beginnings and ends of the intervals of the afternoon prayers

used in ancient and medieval Europe and Asia, so it is not too surprising that they found their way into Islamic religious practice.

Determining the times of the prayers by observation is a relatively simple matter of setting up a gnomon and measuring the shadow length or looking from a high vantage point for the top of the sun to pass the horizon. However, in a crowded city where buildings obscure the view of the sky or the horizon, or when the weather is bad and a clear shadow cannot be seen, it is possible to resort to tables for determining the time of prayer. The earliest known set of tables giving prayer times was formulated by al-Khwārizmī during the ninth century. The tables are designed to be used in the city of Baghdad and give the length of the shadow at the time of prayers for every 6° of solar longitude, and the time of day in seasonal hours from the altitude of the sun as observed by an astrolabe or other device.

Much more sophisticated prayer tables are known from medieval Cairo. The inspiration behind these tables was one of the most important Islamic astronomers, Ibn Yūnus. The son of a *hadīth* scholar and historian, Ibn Yūnus lived in the town of Fustat in Egypt until his death in 1009 AD. Aside from a sideline as an acclaimed poet, he was an active astronomer who made many astronomical observations for the Fatimid caliphs. He compiled more than one *zīj* incorporating astronomical parameters that showed great improvements over those of earlier astronomers, and worked extensively on devising astronomical tables for timekeeping. In the biography written by his contemporary al-Musabbihī, Ibn Yūnus is portrayed in terms that sound very much like the modern stereotype of a scatterbrained academic complete with unkempt clothes and an absent-minded nature. Famously, while still in apparent good health he is said to have predicted that he had only another seven days to live. He saw to his affairs, retreated to his home, and recited the *Qur'an* until the seventh day, when his prophecy came true and he died.

Ibn Yūnus's tables for determining the time of the afternoon prayers were included in an extensive work called, appropriately enough, *Very Useful Tables*. His prayer tables list the altitude of the sun at the beginning of the prayer interval for each degree of solar longitude. Elsewhere in his *Very Useful Tables* he provides tables that allow the solar altitude to be converted into the time of day. Observation of the altitude of the sun using and astrolabe or other instrument, or another method of keeping track of time, could then be combined with these tables to determine the times of prayers.

Today prayer times may be found in many newspapers, in Islamic pocket calendars and even on the Internet. One website announces that it provides prayer times for six million cities worldwide at the click of a mouse. Another offers to send a call to prayer by text to your mobile phone. Behind all these mechanisms for alerting people to the time of prayer lie astronomical calculations identical in purpose, if not necessarily in method, to those of many generations of Islamic astronomers stretching back more than a millennium.

The Direction of Prayer

Astronomy played one further role in the practice of the Muslim faith. Each of the five daily prayers should be performed facing the sacred direction towards the Ka'ba in Mecca. The Ka'ba is a religious shrine built some time before the seventh century in the city of Mecca. The *Qur'an* instructs that prayers should be performed towards this building. Accordingly, beginning with the Prophet Muhammad, Muslims have always faced the Ka'ba in their prayers. Islamic law also directs that many ritual acts such as reciting the *Qur'an* and performing the call to prayer be undertaken facing in this sacred direction, known as the *qibla*. Mosques are also aligned along the *qibla*. Today, Muslims can look in specially prepared tables to find the compass direction of the *qibla* for their location, but in medieval times the *qibla* was often determined by astronomical observation.

The earliest Muslims were faced with an enormous problem when it came to deciding the direction of the *qibla*. Of course, if the Ka 'ba was within sight it was easy to stand facing it, but what if it was too far away to see? What, indeed, if one were several hundred miles away, or even a few thousand – how then could one know which direction to face? Some early Muslims simply faced the direction of the major road one followed to reach Mecca. Others preferred to face due south as the Prophet Muhammad had done when he was in the city of Medina, to the north of Mecca. Still others advocated standing in the same direction as one would at one of the walls of the Kaʿba. Since the Kaʿba itself is astronomically orientated towards the rising point of the star Canopus (α Car), then the *qibla* could be assumed to be along or perpendicular to the direction of Canopus's rising. However, Canopus cannot be seen at all geographical latitudes, and so the risings of other heavenly bodies or even wind directions were sometimes used as proxies. These different methods could yield quite different results, and in some medieval cities different mosques could be aligned in noticeably different directions.

Islamic astronomers knew well, however, that simple methods relying on the rising points of stars were no good. Already by the eighth century AD, astronomers were attempting to find the *qibla* through the application of mathematics to the problem. Essentially it comes down to using spherical trigonometry to find the great circle along the Earth's spherical surface between where one is standing and the Kaʿba in Mecca. Some astronomers chose to use approximate methods to determine the great circle, but there is no doubt that an accurate method was developed by the ninth century.

The possession of a method for determining the *qibla* mathematically was not the end of the matter. In order to use the mathematics, you needed to know the relative geographical locations – the latitude and longitude – of Mecca and your current location. Establishing the latitude of a location can be achieved relatively easily using simple astronomical observations. For example, the altitude of the celestial pole about which the stars circle in the sky once every twenty-four hours may be measured using an astrolabe, and this equals the geographical latitude of where you are standing. However, establishing geographical longitudes is much more complicated.

Whereas latitude is measured from a fixed and obvious place, the Earth's equator, there is no fixed reference line from which to measure longitude. The Greenwich meridian has been arbitrarily defined as the line of zero longitude, but while this line is now accepted worldwide, even in quite recent history there have been debates over where the line of zero longitude should fall, and several 'prime meridians' have existed at the same time during certain periods. For example, the Greenwich prime meridian was established in the mid-nineteenth century and accepted in most countries, but France continued to use its own Paris meridian until shortly before the First World War.

Due to the absence of a fixed zero line of longitude only *differences* in longitude can be measured. Today, as we travel east or west around the world, we change our watches when we cross into new time zones. The idea of time zones is a modern convenience. In the ancient and medieval world time was usually defined locally by the sun: midday was when the sun was at its highest point, lying on the south meridian in the sky. However, as we move to the west, the sun will still be rising in the sky at that instant. If we move directly due west, for every degree of longitude we travel the sun will be four minutes of time away from reaching the meridian. In principle, the difference in longitude can be determined if you know what time it is in one location when the sun is on the meridian at another place. But how do you know the time in the other city? The final solution to this problem came with the development by the Englishman John Harrison in the eighteenth century of a clock that kept accurate time even at sea on a rocking boat. In the ancient and medieval world the only accurate clock that could be used was the sky.

Al-Bīrūnī, an eleventh-century scientist, philosopher, historian and councillor, wrote a whole book on determining the geographical coordinates of cities using astronomical phenomena. He described several different astronomical phenomena that can be used to measure longitude distances if they are observed simultaneously at different locations. The best of these are lunar eclipses. The possibility of using simultaneous lunar eclipse observations to determine longitude differences had been known since antiquity. The principle is simple. As the shadow that causes a lunar eclipse falls on the moon itself, it is seen at the same time all over the world. Therefore, if the time of the beginning of the eclipse measured from midday is known from two different cities, the difference in longitude between the two can readily be calculated by dividing the difference in minutes by four to give the longitude difference in degrees. Al-Bīrūnī went on to criticize a certain al-Hirawī for carelessly writing that solar rather than lunar eclipses should be used to determine longitude differences. As Al-Bīrūnī knew well, because solar eclipses are caused by the moon's shadow sweeping across on the Earth's surface, different observers at different locations will see different eclipses.

Al-Bīrūnī's book is an example of the kind of work done by many astronomers in establishing the geography of the known world in order to allow the *qibla* to be determined for people in different locations. Many astronomers made observations with the aim of fixing geographical location. The end product of all this work was the production of tables containing the *qibla*. Almost every *zīj* contained *qibla* tables. Although many legal scholars preferred the simple methods for determining the *qibla* such as using the direction of the rising of Canopus, some medieval mosques are aligned quite precisely on the theoretical *qibla* as determined by astronomers.

Astronomy held an integral role in the practice of the Islamic faith in the medieval period. It was used to define the calendar, to determine the times of prayer and to establish the sacred direction in which to pray. All three uses demonstrate the practical utility of astronomy in Islamic society. But astronomers are people, and often educated and inquisitive people at that, so it would be wrong to think that they were interested only in the utilitarian aspects of astronomy. The next chapters describe how astronomy was also studied as an intellectual pursuit, with no direct application (astrology aside) than to provide a better understanding of the universe around us.

Astronomical Observations and Instruments in the Medieval Islamic World

Observing astronomical phenomena may have played an important role in everyday religious practice through establishing the times of prayer and the direction of the *qibla*, as described in the previous chapter, but the sky was also observed for other many other reasons. Astronomers made careful observations of planetary positions and lunar and solar eclipses to test and improve astronomical theories. In order to undertake this task, Islamic astronomers designed and built accurate astronomical instruments for measuring celestial positions. Observatories were commissioned and built under the patronage of local rulers and staffed with teams of astronomers set the task of improving astronomical tables.

The sky was seen by every member of society, not only the astronomers, and so all kinds of people looked at the night sky. Many accounts of unusual astronomical events such as bright comets, new stars (e.g. the supernova of 1006 AD), and eclipses of the sun and moon were recorded in chronicles and other historical and literary works. The people who saw these events rarely made careful observations or recorded detailed accounts of what they saw, but their reports tell us a great deal about how common people related to the sky.

Islamic astronomers have rightly gained fame for the astronomical instruments they designed. Chief among these instruments was the astrolabe, which was not only a device for observing the altitude of an object in the heavens, but also an analogue computer that could turn this altitude into the time of day or night and do much more besides. Many hundreds of astronomical instruments made by medieval Islamic astronomers are preserved today in museums around the world.

Astronomical Observations in Medieval Arabic Chronicles

Historical chronicles written by Arabic historians from the ninth to the fifteenth century AD contain a scattering of reports of astronomical events. As would be expected from a historical document, these accounts are mixed in with other happenings: local and national politics, weather, natural phenomena such as earthquakes, and much more. Generally, the astronomical events reported are those that can be noticed easily simply by looking up at the sky: eclipses of the sun and moon, comets, meteor showers and new stars.

Only a small percentage of medieval Arabic chronicles have been printed and studied. The remainder are to be found in manuscripts scattered in libraries throughout the Middle East and beyond. It is therefore very likely that new discoveries of reports of interesting astronomical phenomena will be found in the future.

The most common astronomical phenomena reported in Arabic chronicles are eclipses of the sun and moon. More than seventy reports of lunar eclipses (*khusūf al-qamar* 'failing of (the light of) the moon') and some twenty-seven accounts of solar eclipses (*kusūf al-shams* 'cut (into) the sun') have been identified in a thorough search of printed chronicles by Said S. Said, F. Richard Stephenson and Wafiq Rada of the University of Durham. Many of the accounts of the eclipses are fairly brief, but some include interesting details. For example, the chronicler Ibn Hayyān from Cordoba reports that in 229th year of the Hijra (912 AD),

> ... the sun was eclipsed and it disappeared totally on Wednesday when one night remained to the completion of Shawwāl. The stars appeared and darkness covered the horizon. Thinking it was sunset, most of the people went for the sunset prayer. Afterwards, the darkness cleared and the sun reappeared for half an hour and then set.

On the occasion of this eclipse, which took place on 17 June 912 AD, the sun was totally eclipsed at Cordoba moments before sunset. During a total eclipse the sky can literally appear as if it were night with the full array of stars visible. During the initial stages of the eclipse, the loss of daylight is almost zero and darkness falls at totality almost without warning. It is not surprising, therefore, that as the eclipse had not been predicted in advance, it could easily appear that the sun had set and the night had started. It must have been quite a shock when, after a few minutes, the sun emerged once more from the eclipse, only to set a few minutes later.

Ibn al-Jawzī, a chronicler from Baghdad who lived during the twelfth century AD, quotes an eyewitness account of a solar eclipse that took place on 20 June 1061 AD:

> On Wednesday, when two nights remained to the completion of the Jumāda al-Aula, two hours after daybreak, the sun was eclipsed totally. There was darkness and the birds fell whilst flying. The astrologers claimed that one-sixth of the sun should have remained (uneclipsed) but nothing of it did so. The sun reappeared after four hours and a fraction (of an hour). The

eclipse was not in the whole of the sun (i.e. it was not total) in places other than Baghdad and its provinces.

The description of birds falling from the sky may seem fanciful, but the phenomenon has also been noted by European chroniclers during other total eclipses. This eclipse had obviously been predicted in advance, but Ibn al-Jawzī points out, rather uncharitably, that those making the prediction, presumably from astronomical tables, incorrectly predicted the size of the eclipse.

An account of the eclipse on 12 February 1431 AD by al-Maqrīzī further illustrates the problems faced by the astronomers in predicting eclipses. In this case it had been announced in Cairo that the sun would be eclipsed and people should pray and do good deeds. However, when people watched for the eclipse it did not come. The people who had made the prediction were first denounced for giving a false warning. But then, a while later, news came that the eclipse had been seen in al-Andalus (southern Spain). While this cannot have been a complete vindication of the astronomers who made the prediction, it must have gone some way towards restoring their reputations.

People also made astrological forecasts because of eclipses. For example, the Cairo chronicler Ibn Iyas reported the following story about the lunar eclipse on 8 February 1487 AD:

> In the month of Safar, the moon's body was eclipsed and the earth was darkened. It remained in eclipse for about 50° (3⅓ hours). People were saying that the demise of the Sultan was becoming near. Nothing of what they said happened and the Sultan stayed (in power) for a long time after that.

In fact the Sultan, al-Malik al-Ashraf Abū 'l-Nasr Qaytabā lived for another nine years after the eclipse.

Several reports of the appearance of a new star in 1006 AD are recorded in chronicles throughout the Islamic world. For example, Ibn al-Jawzī, whose report of the eclipse of 1061 AD is quoted above, reported that a large star similar in size to Venus appeared in the sky somewhere to the left of the *qibla* direction on the night of 2 May 1006 AD. He compared the rays of the star to those of the moon shining down on the Earth. According to a Christian scholar in Antioch named Bar Hebraeus, the star was visible for four months before disappearing. This new star was a supernova explosion – the tearing apart of a star producing an extremely bright object in the sky that flares for a few months before disappearing. The remnant of the explosion in 1006 AD has been detected by radio telescopes and is catalogued as G327.6+14.6 by modern astronomers.

Quite a number of comets (*najm dhu dhu'aba*, 'star with locks of hair') are reported in chronicles. Following ancient Greek tradition, comets were considered to be atmospheric phenomena not astronomical events. But comets still had major astrological importance, frequently portending evil events. For example, a comet was reported in a Syrian chronicle for 754 AD like a 'javelin of fire' and remained in the sky until the end of the war between the caliph and his uncle.

Astronomical Observations made by Islamic Astronomers

Surprisingly few reports of astronomical observations by medieval Islamic astronomers are known. In part this is due to the vast amount of unstudied Arabic astronomical manuscripts that remain in collections throughout the world, but it also appears that many medieval astronomers did not consider it necessary to preserve their observations by including them in their works for distribution. The observations were made and used to adjust and improve astronomical theories, but once a set of new parameters had been deduced, the observations appear to have been considered worthless. There are exceptions to this trend, however. Ibn Yūnus published reports of observations of thirty eclipses of the sun and moon made by al-Māhānī, the Banū Amājūr family in Baghdad and by himself in Cairo in his *Hakimī Zīj*.

Ibn Yūnus's motivation for observing and recording eclipses was first to test current astronomical tables, and second to provide empirical data from which to derive new parameters to improve the tables. It was well known in the ancient and medieval world that the calculation of eclipses is a very sensitive procedure. A small error in lunar or solar theory can have a large impact on the accuracy of an eclipse prediction. Eclipses therefore provided a key test for astronomers.

The Islamic astronomers who observed eclipses generally recorded the size of an eclipse and the time it began, its midpoint and its ending, plus the beginning and end of totality for total eclipses. The exact moment of the beginning of a lunar eclipse is hard to spot exactly because the Earth's shadow has a fuzzy edge. This problem was known to Islamic astronomers. For example, the celebrated eleventh-century polymath al-Bīrūnī, who wrote an incredible 146 books, said of lunar eclipses that they do 'not become perceptible to the observer until the segment removed from it (i.e. the moon) according to some authors reaches a limit of one digit' or one-twelfth of the moon's disc. He continued by discussing how to observe solar eclipses. Because the sun's rays are so bright, he said, it is painful to look at the sun and so it is better to look at the reflection of the eclipse in water. As a final warning to those who would observe the sun directly, he lamented, 'My eyesight has been weakened by such observations of solar eclipses in my youth.'

Al-Bīrūnī's suggestion to view solar eclipses by reflection in water seems to reflect the observing practice of many astronomers. For example, Ibn Yūnus quotes an observation by 'Ali ibn Amājūr of the solar eclipse on 18 August 928 AD where a pool of water was used to view the sun:

> I ('Ali ibn Amājūr) observed this eclipse with my son Abū al-Hasan and Muflih and (found) that the sun rose (already) eclipsed by less than ¼ of its surface. The eclipse continued to increase by an amount that we could perceive until ¼ (of its surface) was eclipsed. We observed the sun distinctly (by reflection) in water. (We found that) it cleared and nothing of the eclipse remained and we distinguished the (full) circle of the sun's body in water; (that was) when the altitude (of the sun) was 12° in the east, less ⅓ of a division of the (instrument) *al-halaqah* (i.e. the ring), which is graduated in thirds (of a degree), that is (less by) ⅓°.

Aside from eclipses, most of the observations made by medieval Islamic astronomers focused on the positions of the sun, moon and planets at specific times in order to test and refine astronomical theories. However, not all changes to the parameters underlying a theory were based upon observation. A certain Mu'ayyad al-Dīn al-'Urdī complained in a work written in the thirteenth century that some of the authors of *zījes* made small changes to established parameters to give the illusion that they had made new observations. A similar practice is known from medieval China. There were, however, other astronomers who did make observations.

In addition to observations of the positions of the sun, moon and planets, some astronomers undertook systematic measurements of the positions of the stars. The standard star catalogue available in the medieval period was the list of coordinates and magnitudes of the stars in forty-eight constellations found in Ptolemy's *Almagest*. The tenth-century Persian astronomer al-Sūfī was the first person to attempt to update Ptolemy's catalogue by remeasuring the magnitudes and some of the positions of the stars. He published a famous book, the

13. A page from Al-Sūfī's book describing the constellations.

Kitāb suwar al-kawākib al-thābita ('Book on the constellations of the fixed stars'), which listed the stars in each constellation with their positions and magnitudes and included illustrations of the constellations (see figure 13). Al-Sūfī's book was well known both in the Islamic world and in Europe, where his name was Latinized as Azophi.

For the next four centuries, al-Sūfī's book provided the main source of star positions for Islamic astronomers. These positions were important, as planetary observations were often made with respect to nearby stars, so that the catalogue of stars was essential for converting the planetary observations into latitude and longitude coordinates that were necessary for deriving parameters for planetary theories. It was not until the fifteenth century that another programme of systematic measurement of the positions of stars was undertaken. Ulugh Beg, the grandson of Timur the Tarter and ruler of Mawaraunnahr in central Asia, was an active astronomer as well as ruler. His name means 'great prince' and was adopted by Ulugh Beg, formally Muhammad Taragay, while he was still a young man. In 1420 AD Ulugh Beg founded a *madrasa* (a school of higher education) in Samarkhand, selecting the scientists who were employed to teach there himself. Among these scholars was al-Kāshī, an expert astronomer. Together, Ulugh Beg, al-Kāshī and the employees at the observatory in 1424 AD undertook the production of a new catalogue containing 1,018 stars with newly measured positions where possible (a few star positions were taken from al-Sūfī's book).

Astronomical Instruments

Many reports of astronomical observations made by medieval Islamic astronomers indicate that various instruments were used during observations. For example, Ibn Yūnus reports that the astronomer al-Māhānī in Baghdad used an astrolabe to measure the altitude of Aldebaran (a bright star in Taurus) at the moment when the lunar eclipse on 22 June 856 AD began. From the star's altitude al-Mahanī derived the time of the eclipse. The refinement, construction and development of fine astronomical instruments provide one of the most visible illustrations of the achievements of medieval Islamic astronomers.

Many hundreds of astronomical instruments are preserved from the Islamic world. These include celestial globes, sundials and quadrants, but without question the king of Islamic instruments was the astrolabe. About four times as many astrolabes are preserved as any other type of instrument, and they are the instruments most frequently referred to in reports of observations. When astronomers appear in Islamic art they are very often pictured with their astrolabes (see figure 14).

The astrolabe has dual uses within astronomy. When an astrolabe is hung by the ring at its top and allowed to settle freely, its back can be used to measure the altitude of a celestial body using the moveable alidade (a sighting bar free to rotate around the centre of the astrolabe) and a scale written towards the edge of the instrument (see figure 15b). The alidade would often be equipped with either sighting holes or bars to allow the alidade to be accurately aligned with a heavenly body. Once this had been done, the altitude of the sun, moon, planet or star could simply be read off.

When an astrolabe is turned over to reveal its front, a plate inscribed with curves and lines can be seen. Attached to the plate by a central pin is a moveable skeletal part called a 'rete' or 'spider' (see figure 15a). The front side of the astrolabe functions as a mechanical computer allowing the time to be determined from the altitude or direction of a heavenly body and vice versa. The curves inscribed on the plate are a projection of the three-dimensional sky onto a two-dimensional surface (see figure 16). The vertical line corresponds to the zenith meridian running from north to south over an observer's head. At a point on this line determined from the geographical latitude of the

14. Nasīr al-Dīn al-Tūsī and colleagues at the Maragha observatory.

observer, the midpoint of heaven is marked, and surrounding this point are concentric curves corresponding to the points of equal altitude. The lowest of these altitude circles is the horizon. Marked along the horizon are intervals of azimuth running from southeast on the left of the astrolabe to southwest on its right. Curves extend from these points up to the point of mid-heaven.

The rete is made up of a metal circle fixed off-centre from the pin with a metal web of points inside and outside this circle (see figure 17). The

15. A brass astrolabe made by 'Abd al-Kaīm al-Misrī, (a) obverse, (b) reverse

circle corresponds to the ecliptic and is divided into the twelve zodiacal signs, each containing 30°. The other points on the rete mark the positions of certain bright stars. Finally, the rete contains a pointer onto a scale marked on the edge of the astrolabe. This scale corresponds to the 360° of the celestial equator – the 360° daily

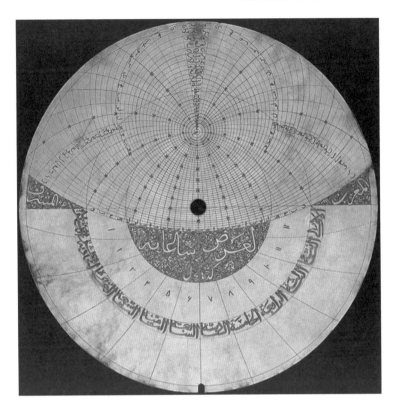

16. One of the plates from a seventeenth-century Persian astrolabe

rotation of the sky as seen from the Earth.

By moving the rete and reading the position of the pointer on the scale, it is possible to use the astrolabe to measure intervals of time. For example, if the sun is measured to be at a certain altitude, you simply rotate the rete until the sun's position along the ecliptic is placed on the appropriate altitude circle on the plate. The sun's position can be calculated either accurately by tables or approximately simply by assuming that the sun moves through the zodiac at about 1° per day and counting from the date of the spring equinox when the

17.   The rete from a seventeenth-century Persian astrolabe

sun is at the beginning of Aries. Once the sun's position has been deduced and the position of the pointer when the sun is at the correct altitude has been noted, the rete can be rotated until the sun's position in the ecliptic is now on the vertical meridian on the plate and the pointer's new position read. Since the sun will be on the meridian at noon, the difference between the two noted pointer positions equals the time before or after noon given in degrees of time (360ths of a day or 4 minutes). The same process can be used to determine the time from the altitude of one of the stars marked on the rete. Additionally, the azimuth of the sun or a star can be determined at any time by aligning the pointer to the appropriate position corresponding to the number of degrees before or after noon and reading off the sun or star's position against the azimuth curves. The astrolabe can therefore be used as a clock, compass and computing device, as well as an observational tool.

Underlying the construction of an astrolabe is a mathematical theory for the projection of the sky onto the two-dimensional surface. A method for making this projection was proposed by ancient Greek astronomers, and it seems quite likely that astrolabes were used by some ancient astronomers. However, medieval Islamic instrument makers made the device their own.

Several Islamic astronomers and mathematicians wrote treatises on the construction of the astrolabe. One of the best known, and also one of the earliest, was by al-Farghānī. Early in his life Ahmad ibn Muhammad ibn Kathīr al-Farghānī was employed by the Abbasid caliph al-Ma'mūn, who reigned in Baghdad from 813 to 833 AD. His best-known work was a thirty-chapter overview of Ptolemaic astronomy omitting all mathematical discussion, written some time after al-Ma'mūn had died but before he wrote his book on the astrolabe, which was probably published in 856 or 857 AD. His book on Ptolemy's *Almagest* was known throughout the Islamic world and in Europe, where two Latin translations were made in the twelfth century by John of Seville and Gerard of Cremona. The caliph al-Mutwawakkil later engaged al-Farghānī to supervise the construction of the 'Great Nilometer' in Old Cairo and the digging of a canal from the Tigris river. But according to a story reported by Ibn Abī Usaybi'a, al-Farghānī's contribution to engineering projects was not always positive: he made the beginning of the canal too deep so that the water flow along the canal was too low unless the level of the Tigris was very high. Fortunately, another engineer, a certain Sanad ibn 'Alī, who had himself been overlooked for the job, confirmed al-Farghānī's calculations, and as the caliph died shortly before the canal was completed, al-Farghānī escaped punishment.

Whereas al-Farghānī's overview of Ptolemaic astronomy had been purely descriptive, his book on the astrolabe is a detailed mathematical description of the instrument and its construction. He computed many tables for the construction of astrolabe plates relevant for different latitudes, and explained how the curves were to be drawn on the plates. He ended the book with a somewhat ill-tempered chapter explaining that an astrolabe could only be constructed in the way he had described, despite the 'absurd' opinions of others.

Many Islamic astrolabes are objects of beauty as well as scientific instruments. Some are highly decorated with fine calligraphy marking the names of the zodiacal signs and the stars. Very often the rete incorporated artistic representations of animals with the markers for stars built into the design. It is no wonder that astrolabes were prized possessions in the medieval world and that their makers were praised by such important scientists as al-Bīrūnī.

Astronomical Observatories in the Islamic World

The Abbasid caliph al-Ma'mūn established a tradition of patronage of astronomy through the building of observatories. Astronomers were employed in observatories to undertake programmes of astronomical observations, update the parameters of earlier astronomical theories (or sometimes update the theories themselves) and compile astronomical tables. Al-Ma'mūn became caliph in 813 AD following a civil war between him and his brother, who had been named heir to the caliphate by their father. Following his elevation to caliph, he patronized the study of science and the humanities. He brought scholars from throughout the caliphate to Baghdad, irrespective of their religion, and encouraged them to interact with the 'House of Wisdom', formally an extensive library and centre of translation of

Greek and other scientific works into Arabic, but also a kind of learned society for the study of science.

In about 828 AD, al-Ma'mūn founded an astronomical observatory in Baghdad, and another a few years later in Damascus. He engaged several astronomers to make daily observations of the sun and moon in order to test and correct existing astronomical tables. To aid the astronomers in their work he also commissioned the construction of many astronomical instruments. The results of the observations made in Baghdad under his patronage were incorporated into a new set of astronomical tables known as the *Mumtahan Zīj*, meaning 'tested tables'.

Among the astronomers who benefited from al-Ma'mūn's patronage was al-Khwārizmi, although it is not known if he worked at one of the observatories. As well as producing highly influential work on algebra, with which his name will forever be linked, he compiled a *zīj* (a set of astronomical tables with explanatory text) in Baghdad. This combined elements of Indian, Persian and ancient Greek astronomy and despite criticism from other astronomers, notably al-Farghānī, al-Khwārizmi's *zīj* was quickly distributed throughout the Islamic world and was even eventually translated into Latin in the twelfth century. Although he made observations to determine the obliquity of the ecliptic for al-Ma'mūn, he stuck with the much less accurate value found in earlier astronomical works when constructing his *zīj*.

Both al-Ma'mūn's Baghdad and Damascus observatories had short lives, operating for only a couple of years. In the following centuries several other observatories were set up throughout the Islamic world. Many of these followed al-Ma'mūn's model of the observatory as a specialist institution with staff dedicated to astronomical observation and research. Most of the institutions were created with the support of the caliphs or members of royal families.

Undoubtedly the most important observatory in the medieval Islamic world was the Maragha observatory in northern Iran. The observatory was set up during the reign of the Mongol ruler Hulegu, who had defeated the Abbasid caliphate. Hulegu appointed Nasīr al-Dīn al-Tūsī, an astronomer and adviser to Hulegu, to form and run the observatory, and building work began in the late spring of 1259 AD. The construction of the observatory took several years, but on its completion al-Tūsī had founded the largest and most well-equipped observatory that had then been known in the Islamic world.

The observatory complex was situated on top of a hill close to the city of Maragha. In addition to the main building there was a domed building that served as a solar observatory, a library reported to have contained 400,000 works (perhaps located within the main observatory building) and several instruments placed out of doors for making observations. Most of the instruments for the observatory were made by Mu'ayyad al-Din al-Urdī, a Syrian astronomer. Al-Urdi wrote a treatise describing the instruments he built for the observatory. These include an armillary sphere, solstitial and equinoctial armilla, an azimuth ring and a parallactic ruler, as well as specially designed instruments such as a device with two circular holes used to measure the apparent diameters of the sun and the moon and a large mural quadrant with a radius greater than 4 metres. An even bigger mural quadrant was later build at the observatory of Ulugh Beg in Samarkhand. This large instrument was built into a channel dug in the earth and had a radius of 40 metres.

A pioneering feature of the Maragha observatory was its role as a teaching institution as well as a place of astronomical observation and research. At times up to a hundred students received instruction in astronomy and mathematics at the observatory. It is not surprising that it was in this environment that perhaps the greatest contributions to the development of theoretical astronomy by Islamic astronomers were made. As discussed in Chapter 7, Nasīr al-Dīn al-Tūsī and his associates and successors at the observatory proposed ingenious reforms to several problematic aspects of Ptolemaic astronomy that would have important resonances in the eventual replacement of Ptolemy's astronomy during the Renaissance.

Medieval Planetary Theories

During the early years of Islam astronomers combined pre-Islamic folk astronomy from the Arab lands with astronomical knowledge found in the countries conquered by Islamic rulers. Persia provided a link between Indian astronomy and the Islamic world, and many early Islamic astronomers learned the techniques of Indian astronomy. This had developed out of a combination of the Babylonian and Greek traditions, but had grown into a science of its own through the application of new mathematical techniques invented in India in the early part of the first millennium AD.

By the eighth and ninth centuries AD, Greek astronomical texts had found their way into the hands of Islamic astronomers. In particular, Ptolemy's *Almagest* was discovered by Islamic astronomers, and translations of the book into Syriac and Arabic were produced. Ptolemy's astronomy soon became the basis of almost all studies into theoretical astronomy by Islamic astronomers. That is not to say that the Islamic astronomers simply blindly followed Ptolemy's astronomy. Many astronomers attempted to correct what were seen as flaws in Ptolemy's work by making new observations to refine Ptolemy's parameters, or by developing new geometrical techniques to replace parts of his models.

Along with Indian and Greek astronomy, Islamic astronomers were also exposed to astrology. Moreover, just as in the Christian world, astrology would be both embraced and ostracized. Most theologians condemned astrology on the grounds that nobody can know the future but God. Other people criticized astrology's lack of utility, arguing that it was a pseudo-science and that the predictions made by astrologers very rarely came true. One critic, al-Tabarī, told the story that on his deathbed the ninth-century caliph al-Wathiq gathered seven astrologers to foretell his future. They all agreed that the caliph would recover and go on to rule for another fifty years, but five days later the caliph was dead and the astrologers' reputations were in tatters. Despite stories such as this, astrologers were always able to point to what they claimed were successful predictions, and their claims were usually met by favourable responses from the general population and their rulers.

The most widely used form of astrology in the Islamic world was the casting of horoscopes. The astronomical data needed for a horoscope usually included the positions of the sun, moon and planets in the zodiac and the *horoscopus* point – the point of the ecliptic rising over the eastern horizon at the moment the subject was born. The only practical way of obtaining this data was by using astronomical tables – searching through old observation reports for the relevant data would be both time consuming and often fruitless on account of bad weather. The needs of astrologers for good astronomical tables played an important role in the promotion of astronomy to potential patrons. When the cost of the construction of the Maragha observatory was put before the ruler Hulegu, for example, Nasīr al-Dīn al-Tūsī enticed Hulegu into providing the funds by pointing to the usefulness of astrology to his empire.

The boundary between astronomy and astrology was extremely blurred, indeed almost non-existent, in the medieval period. In their classifications of the sciences, al-Fārābī and later Ibn Khaldūn both included astrology among the sciences of the heavens. Almost all astronomers also cast horoscopes and wrote on the subject of astrology. For example, al-Bīrūnī, one of the greatest scientists of the Islamic world, wrote a long book entitled *Kitāb al-tafhīm li-awā'il sinā'at al-tanjīm*, an instruction manual for astrology. The first two-thirds of the book provides an introduction to the astronomy that is necessary to understand the astrology discussed in the final part.

Astrology was not the only reason for studying astronomy. The practical utility of astronomy to everyday life also encouraged the development of the science. A few individuals also studied astronomy for its own ends. As discussed in this chapter, some astronomers dedicated themselves to improving the mathematical, philosophical and physical basis of theoretical astronomy. Most of these improvements had little impact on the accuracy of astronomical tables, and were therefore not important for astrologers. However, they mark an important stage in the history of the science of astronomy.

Islamic Astronomical Tables

All Islamic astronomers probably used astronomical tables to calculate planetary positions, the visibility of the moon or the date and visibility of a lunar or solar eclipse. The tables they consulted would frequently be found in a document called a '*zīj*'. According to the *Encyclopaedia of Islam*, the standard English-language reference work of Arabic studies, the term *zīj* refers to 'an astronomical handbook with tables'. Most well-known Islamic astronomers, and several less well-known individuals as well, compiled *zījes*. Many run to several hundred pages and contain extensive sets of astronomical tables accompanied by explanatory text. The texts range from brief instructions for using the tables to detailed discussions of the astronomical models that underlie the tables, the parameters used in the theories and the observations that were used in deriving the parameters. Around 200 separate *zījes* are now known to have been compiled, although many have been lost or remain unidentified among the mass of unpublished Arabic astronomical manuscripts scattered throughout the world's libraries. Not every one of the 200 *zījes* is based upon original work, however. Many were derived from the predecessors by simply tweaking the underlying parameters to make it appear that they contained new work. Islamic astronomers were not alone is this practice – many medieval Chinese astronomical systems show only cosmetic differences from earlier works.

Most *zījes* contain tables for calculating solar, lunar and planetary longitudes; lunar and planetary latitudes; solar and lunar eclipses; lunar and planetary visibility; and the *qibla*. In some cases more than one set of tables is required in the calculation. For example, for each of the sun, moon and planets, one table provides the longitude of the body assuming it moves uniformly with its mean motion, and other tables provide corrections to this preliminary position to take into account the body's variable velocity. This approach of first calculating a preliminary position that is then modified to give the true longitude was adopted by Islamic astronomers from the example of Ptolemy's *Almagest*. It has echoes of the Babylonian method of calculating the length of the lunar month by first assuming that the variation is purely due to the variable velocity of the moon and then applying a correction for the solar anomaly. The tables also follow the Greek tradition of writing numbers using sexagesimal notation, a system that unbeknown to the medieval Islamic astronomers had its origins in ancient Mesopotamia.

A new development by Islamic astronomers in the construction of astronomical tables was the replacement of the rather clumsy Greek 'chord' function with a sexagesimal version of the 'sine' function developed by Indian mathematicians and still used today. One of the great achievements of many of the authors of *zijes* was to construct extensive trigonometric tables. The most impressive of these were included in the *zij* of Ulugh Beg and contain sine and tangent values, some of which are accurate to the fifth sexagesimal place for each minute of arc. This was a truly mammoth enterprise that required calculating values for more than 5,000 entries.

'Doubts concerning Ptolemy'

Following the translation of Ptolemy's *Almagest* into Arabic in the early part of the ninth century AD, Islamic astronomers almost universally adopted Ptolemy's astronomy as the basis for constructing *zijes*. The astronomical theories Ptolemy proposed in the *Almagest* were held to be fundamentally correct and any problems in the accuracy of the predictions based upon these theories were thought to be due to the accuracy of the parameters Ptolemy adopted. Because Ptolemy only had access to scattered observations stretching over a few centuries, it was not surprising that some of the parameters he derived from them were inadequate. All that was needed was to correct these parameters by making new observations and use Ptolemy's methods to derive the parameters for the theories afresh. Moreover, because of the long time period between these observations and those listed by Ptolemy, one would inevitably get better values. One of the foremost astronomers of the ninth century, Habash al-Hāsib, made just this point. All the ancient astronomers have left for us to do, he argued, is to correct mistakes in the parameters using the methods described by those ancient astronomers.

One exception to this general acceptance of astronomy came about through observation of the path of the sun and the fixed stars. By simple measurements of the altitude of the sun at noon on the days of the summer and winter solstices it is easy to determine the angle of the obliquity of the ecliptic – the angle between the great circle along which the sun moves and the celestial equator. Astronomers in the service of al-Ma'mūn measured the obliquity of the ecliptic as 23;33° whereas according to Ptolemy the angle was 23;51,20°. Similarly, comparison of the celestial longitudes of stars measured by Islamic astronomers with the positions given in the star catalogue in Ptolemy's *Almagest* showed that they had changed by about 1° in 65 years, whereas according to Ptolemy the rate of change, known as 'precession', is about 1° in 100 years. How had Ptolemy got these figures wrong?

The answer put forward by most ninth-century Islamic astronomers was partly wrong and partly right. Ptolemy had not got the measurements wrong, they said, but the rate of precession and the obliquity of the ecliptic must be slowly changing – too slowly for Ptolemy to have been able to detect from the relatively recent observations available to him, but now seen because of the extra 800 years that had elapsed since Ptolemy's time. In fact, although the obliquity of the ecliptic does change slowly over time, albeit at a slower rate than was claimed, the rate of precession does not change: Ptolemy simply got the number wrong by about 30 per cent.

A theory to explain the changes in the obliquity and the rate of precession was put forward in the ninth century by Thābat ibn Qurra. He was born in Harran in *circa* 824 AD, but spent most of his life in Baghdad, where he wrote several works on mathematical subjects and worked as a physician. He practised the Sabian religion, which included worship of the stars, and it is tempting – but almost certainly wrong – to link this with his interest in astronomy. His theory, known as 'trepidation', was inspired by a strange comment by the fourth-century Greek astronomer Theon of Alexandria, who wrote several commentaries on Ptolemy's astronomical works. In Theon's commentary to Ptolemy's *Handy Tables*, he mentioned that some astrologers maintained that precession was cyclical: the stars move to the east at a rate of 1° every 80 years until they have moved by 8°, at which point they travel back at the same rate but in the opposite direction until they have returned to their starting point, when the cycle starts again. Thabat transformed this cyclical idea into a system compatible with Ptolemy's astronomy. He added an extra ninth sphere to the Aristotelian universe to account for the daily rotation of the sky, and gave to the sphere of the fixed stars (the eighth sphere) its own motion. The motion of this eighth sphere caused a slow cycle in the obliquity and the rate of precession, solving two problems with one simple modification to Ptolemy's system.

Not all astronomers accepted Thābat's theory of trepidation. Few supported it in the eastern Islamic world. Al-Battānī, a younger contemporary of Thābat also born in Harran, rejected the idea and adopted a constant rate of precession of 1° in 66 years. Similarly, Ibn Yūnus took the rate of precession to be constant as 1° in 70 years. In Islamic Spain, however, trepidation found its way into many astronomical tables. The so-called *Toledan Tables*, an enormously popular medieval set of astronomical tables based upon a mishmash of astronomy from Ptolemy, al-Khwārizmī and others, included trepidation tables. Even when it became clear by the thirteenth century that a cyclical to-and-fro movement of the stars contradicted observations, some medieval astronomers tried to couple the theory of trepidation with an additional constant forward motion of the stars to produce a varying but always positive rate of precession. Such ideas were dismissed as ridiculous and, more importantly, were contradicted by careful observations by most Islamic astronomers.

It was not until the eleventh century that the whole basis of Ptolemy's astronomical theories was first seriously questioned. The first writer to be openly critical of Ptolemy in this way was a man named al-Hasan Ibn al-Haytham. Born in 965 AD, al-Haytham was an ambitious individual who as a young man had claimed he could build a device on the Nile to regulate the flow of water along the river. According to Ibn al-Qiftī the caliph al-Hākim was so impressed by al-Haytham's proposal that he brought the young man to Egypt to begin the project. However, when al-Haytham travelled around Egypt and discovered that the course of the Nile in southern Egypt did not flow as he expected, he admitted to the caliph that the project was impossible. Although al-Hākim merely transferred him to another office of government, al-Haytham feared for his life after his failure. He came up with an ingenious plan to escape from the caliph: he would pretend to be mad. After successfully persuading those around him that he was deranged, al-Haytham was shut away in his house and remained there until the caliph's death. At that point he suddenly emerged again, proved he was indeed sane and commenced a new life copying scientific texts for a living. He also set about writing his own books.

Al-Haytham wrote more than sixty books on subjects as diverse as ethics, poetry and Aristotle, as well as scientific works on astronomy, mathematics and optics. His main book on optics was translated into Latin around the turn of the thirteenth century and became a widely studied book in Europe, where al-Haytham was known as Alhazen. Al-Haytham's early works on astronomy are not especially remarkable, but in later life he wrote a work that was to have a profound effect on the development of astronomy. Al-Haytham called his book *al-Shukūk 'ala Batlamyūs* ('Doubts concerning Ptolemy').

Early in his career al-Haytham had read the works of Aristotle and become a committed believer in Aristotle's philosophical cosmology. He believed that any astronomical theory must have a physical as well as a mathematical basis. Thus, the circles carrying the sun, moon and planets must be solid spheres. If there were solid spheres they must be subject to the physics of solid bodies. Al-Haytham criticized Ptolemy for breaking this rule. Some of his criticisms were concerned with Ptolemy's imprecise use of language, others with inconsistencies in

Ptolemy's arguments. However, his most serious criticism, which was taken up by later astronomers and required the development of new astronomical models, was of Ptolemy's theory of the planets, and in particular Ptolemy's equant point. Al-Haytham's criticism was simple and direct. If all the circles in Ptolemy's theories are to be interpreted as solid spheres, how can they move so that they do not have constant velocity as seen from the centre of the sphere, but from a point located away from the centre? Imagine a car wheel spinning on its axle. How can different parts of the edge of the wheel move with different speeds? A solid sphere with an equant point is physically impossible. According to al-Haytham, Ptolemy knew this but ignored the problem. For al-Haytham, sweeping the issue under the carpet as Ptolemy had done was not an option. Only one course remained: Ptolemy's model must be rejected.

The problem was that Ptolemy's model worked very well. Ptolemy had introduced the equant into his model because he could not find any other way of obtaining good agreement between the model and observation. So if al-Haytham was right and the equant was impossible, something would have to be found to replace it. This challenge led to some of the most remarkable developments in the history of astronomy.

Nasīr al-Dīn al-Tūsī and the Astronomers of the Maragha School

Born the son of a legal scholar in the city of Tūs in Persia in 1201 AD, Nasīr al-Dīn al-Tūsī had a long and eventful life on the way to becoming one of the most respected scientists in the Islamic world. He studied theology and science under his father and learned philosophy from his uncle before moving to Nīshāpur to partake of a formal education from noted teachers such as Farīd al-Dīn al-Damad. But this was a time of upheaval within Persia. The Mongol empire had ambitions to expand into the Arab lands, so al-Tūsī retreated into the strongholds of the

Isma'ili sect. Here he remained in relative peace and solitude and wrote several important philosophical and scientific works. However, when the Mongol ruler Hulegu conquered the Isma'ili lands in 1256 AD, al-Tūsī could no longer remain in seclusion.

Fortunately for al-Tūsī, Hulegu was interested in astronomy, or more probably astrology, and knowing of al-Tūsī's reputation as a scientist appointed him as a personal adviser. Following Hulegu's capture of Baghdad in 1258 AD, al-Tūsī returned with Hulegu to Persia and was commissioned to set up and run an observatory at Maragha. It was at the Maragha observatory that he and some of his colleagues and students developed highly innovative and ingenious techniques for reforming Ptolemy's theoretical astronomy.

As Ibn al-Haytham had stressed two centuries earlier, the notion of a point some distance from the centre of a circle about which a body on the edge of the circle moved with uniform velocity did not make physical sense. This was only one of the failings of Ptolemy's models for the sun, moon and planets. Simple observations clearly illustrated another problem: according to Ptolemy the distance of the moon from the Earth, and hence also the apparent diameter of the lunar disk, varies hugely during a month – but only a very small variation is seen.

Al-Tūsī developed a mathematical device to overcome some of these problems. Known today as the 'Tūsī-couple', it consists of two circles – spheres if we like – one of which is half the size of the other and rolls around the inside of the larger circle. If the inner circle moves in the opposite direction but at twice the speed to the outer circle, then a point on the inner circle traces out a straight line (see figure 18). In the Aristotelian world view, where everything in the heavens must be described by uniform circular motions, al-Tūsī had succeeded in developing a philosophically correct mechanism for producing linear motion.

The Tūsī-couple was used by al-Tūsī to try to eliminate some of the problems that had been identified in Ptolemy's astronomy. By replacing Ptolemy's 'crank' model for the moon, in which the centre of the deferent circle moved around the Earth on a smaller circle (see above, figure 9) and which caused the unrealistically large variation in the apparent size of the moon, with a Tūsī-couple mechanism within the epicycle, al-Tūsī at a stroke eliminated the problem of the exaggerated change in the moon's size. With the invention and application of this simple mathematical device, he brought about the first significant change in astronomical theory in Europe or the Islamic world for over a thousand years.

Al-Tūsī took a similar approach to Ptolemy's models of the planets. He moved the centre of the deferent circle to Ptolemy's equant point and added a Tūsī-couple device to the epicycle to compensate (see figure 19). Further refinements were proposed in the fourteenth century by Ibn al-Shātir, an enigmatic but extremely significant figure in the history of astronomy.

Ibn al-Shātir was born in Damascus and spent most of his life there.

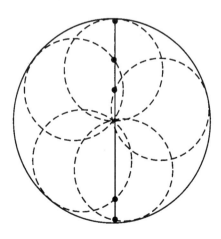

18. The Tūsī-couple

Following his father's death and a brief period living in his grandfather's household, he travelled to Egypt while still a child of around ten years of age to study astronomy. On the completion of his studies he returned to Damascus, and in time became head *muwaqqit* at the Umayyad mosque. Almost nothing else is known about his life, not even his date of death.

During his time working at the Umayyad mosque, Ibn al-Shatir built several astronomical instruments. Among these were an astrolabe, still preserved and currently located in the collection of the Observatoire de Paris, and two new forms of quadrants, which he described in several treatises. His most impressive instrument, however, was a sundial mounted on the northern minaret of the mosque. The sundial was made from a large slab of marble inscribed with several curves systematically arranged to allow the time of day to be read in equinoctial hours throughout the year and the time of the afternoon prayer. Sadly, the original instrument was damaged in the nineteenth century by the incumbent *muqaqqit*, an unfortunate man by the name of al-Tantāwī who was in the process of making some adjustments to the instrument. He attempted to redeem himself by replacing the instrument with a replica that can still be seen at the mosque today.

Ibn al-Shātir's most important contribution to the history of astronomy, however, was his development of new planetary models. He took al-Tūsī's proposals for remedying the defects in Ptolemy's models further to produce theories for the moon and planets that were both accurate and philosophically acceptable. Not only did he reject the idea of the equant, but he also wanted to remove the necessity for eccentric spheres and to return the Earth to its rightful place at the centre of the model. He described his solution to this problem is a treatise entitled *Nihāyat al-sūl fī tashīh al-usūl* ('A final study concerning the correction of planetary theory').

In developing his astronomical theories Ibn al-Shātir made extensive

19. Naṣīr al-Dīn al-Ṭūsī's planetary model.

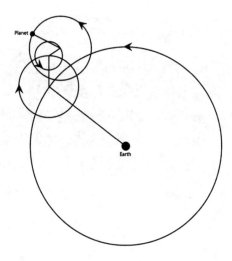

20.   Ibn al-Shātir's model for an outer planet

use of observations he had made of the sun, moon and planets. These observations played an important role for they not only allowed him to deduce new parameters, rather than having to convert those of Ptolemy or earlier Islamic astronomers into his new system, but also provided the motivation behind some of Ibn al-Shātir's changes to earlier astronomical theories. He not only objected to some of Ptolemy's models on philosophical grounds, but also had observational evidence that proved the models needed improving.

At the heart of Ibn al-Shātir's development of improved astronomical models was the replacement of eccentric circles by epicycles, and the replacement of the equant point by yet more epicycles. For example, his model for the outer planets has an epicycle on an epicycle on an epicycle on a deferent circle (see figure 20)! The model may look complicated and deriving the relative sizes and rotation speeds of each epicycle was no easy task, but the end result was both accurate and elegant.

Ibn al-Shātir's work represents the highest point in the development of planetary and lunar models by Islamic astronomers. The decline was swift and unending. By the early part of the fifteenth century, astronomers had forgotten or put aside their qualms concerning Ptolemy. The *zījes* produced in Samarkhand by al-Kāshī and Ulugh Beg, for example, were prepared using essentially Ptolemaic planetary theory in the form used in early works by al-Tūsī. However, Ibn al-Shātir's work did somehow become known in Renaissance Europe.

Recent research has suggested that the Viennese astronomer and mathematician Georg Peurbach may have used some of al-Tūsī and Ibn al-Shātir's methods in his own attempts at refining Ptolemy's planetary models. If so, he would have almost certainly passed them on to his student, Johannes Muller of Konigsberg, widely known by his adopted name Regiomontanus. He was one of the finest mathematicians and astronomers of the Renaissance and his mathematical investigations of Ptolemy's *Almagest*, published in his *Epitome of the Almagest*, were used extensively by the much better known Nicholas Copernicus, inventor of the heliocentric hypothesis that removed the Earth from the centre of the universe and put the sun in its place. It is therefore highly suggestive that Copernicus's models for the moon and planets are mathematically identical to those of Ibn al-Shātir, except for the transformation from an Earth-centred to a sun-centred universe. The transformation, while conceptually difficult, was mathematically trivial, nothing more than routine work for a trained mathematician such as Copernicus. The question remains whether Copernicus developed the same system as Ibn al-Shātir independently, or if one of the great figures of European science drew on knowledge of Islamic astronomy. If he did, he did not mention it in his writings.

Legacies

The history of astronomy is more than an account of observation and calculation, of innovation and individuals. It is also a story of the transmission of astronomical knowledge from one generation to the next, and from one culture to another. Astronomy in the Middle East was first developed in ancient Mesopotamia, passed on to India and Greece, then to the Arab lands, and finally to Europe in the late medieval period. Each culture drew upon the legacy of earlier cultures, taking elements of its predecessor's knowledge and assimilating them into its own astronomy, where they would be adapted, corrected (sometimes wrongly), rewritten and developed into something new and distinctive.

Transmission and assimilation of astronomical knowledge may take many forms. Whole systems of astronomy may be taken over, as happened in Oxrhynchus, where Greek astronomers used Babylonian mathematical astronomy, or in the early Islamic period when astronomers relied on Ptolemy's astronomy for producing astronomical tables. Alternatively, observations or numerical parameters may be taken from one culture and used in a different type of astronomy by the recipient culture. For example, the value for the mean length of the synodic month, 29;31,50,8,20 days, taken from the Babylonian System B arithmetical lunar scheme, was used by Hipparchus and Ptolemy in their geometrical lunar theories. The Babylonian value for the mean length of the synodic month also appears in the astronomy underlying the Jewish calendar and even in a numerical appendix explaining the method of Easter-computus integral to late-Medieval and Renaissance Books of Hours. For the authors of these books, 29;31,50,8,20 days was a canonical value and they did not know of its origin in the ancient Near East.

The decline of the scientific 'golden age' of Islam, to use the historian of Islamic astronomy George Saliba's term, and the rise of European astronomy during the Renaissance, did not prevent Islamic astronomy leaving its own legacy. This chapter outlines several aspects of Middle Eastern astronomy that became part of European astronomy, some of which remain today.

Legacies of Islamic Astronomy in Medieval and Renaissance Europe

Two principal routes were available for the transmission of Islamic astronomy into Christian Europe: Spain and Byzantium. Within two centuries of the founding of the Muslim faith, Islam had spread from the Middle East along northern Africa and into much of Spain. Over the next seven centuries Islamic rule in Spain rose and waned before the Muslim king finally surrendered to the Christian armies from the north of Spain in 1492 AD. But during extended parts of the intervening period, relations with Christian Europe were more cordial and many Europeans travelled to southern Spain to study in the Islamic libraries and schools.

Among the astronomical knowledge that passed from the Middle East through Muslim Spain into Europe were astronomical tables, particularly those of al-Khwārizmī, which were translated into Latin by Adelard of Bath, and the design and use of the astrolabe and other instruments. In addition, Islamic astronomers within Spain compiled new sets of astronomical tables, notably a collection of tables based somewhat inconsistently upon the work of al-Khwārizmi and al-Battānī compiled for the longitude of Toledo and known today as the *Toledan Tables*. They also wrote books describing instruments and astronomical and astrological theories. Many of these works were translated into Latin and became widely known in Christian Europe. The *Toledan Tables* in particular were used extensively throughout Europe and were only eventually superseded in the thirteenth century by a set of tables known as the *Alfonsine Tables*, named for Alfonso X of Castile. Alfonso (1221–84 AD) was an important patron of science in the Middle Ages. He commissioned translations of several Arabic astronomical treatises and new works on various instruments used in astronomical observations.

The second route by which Islamic astronomy reached Europe was via Byzantium, which was closer to Damascus, Baghdad and the other centres of learning in the eastern Islamic world. As a result some astronomical works by Islamic astronomers based in these centres became known in Europe. For example, recent research by Charles Burnett, an historian of Islamic influences in western Europe at the Warburg Institute in London, has traced the transmission of a version of al-Sūfī's astronomical tables and an Arabic translation of Ptolemy's *Almagest* through Antioch during the twelfth century.

By the fifteenth and sixteenth century the work of several important Islamic astronomers was known in Europe. Al-Battānī's *zīj al-Sābi* was translated, probably without its tables, by Plato of Tivoli. Al-Farghānī's introduction to astronomy and a similar introduction by al-Qabīsī were translated into Latin and became very influential. Regiomontanus in his book *The Epitome of the Almagest*, an extremely important work describing the mathematical and geometrical basis of Ptolemy's astronomy, referred to al-Battānī and cited some of his observations. Nicholas Copernicus also mentioned al-Battānī frequently and used a few of al-Battānī's observations in his famous book *De Revolutionibus*, in which he proposed his final version of the heliocentric (sun-centred) cosmology, breaking with almost 2,000 years of tradition.

Copernicus had more Islamic astronomy at his disposal. He frequently used the Tūsī-couple, invented by the driving force behind the Maragha observatory Nasīr al-Dīn al-Tūsī in the thirteenth century, and his lunar and planetary models are mathematically identical (except for a transformation from an Earth-centred to a sun-centred universe) to those of the fourteenth-century astronomer from Damascus, Ibn al-Shātir. Copernicus mentioned neither of these astronomers, nor anyone else as the person who first developed these aspects of his astronomy. It is quite possible that the identities of al-Tūsī and Ibn al-Shātir were unknown to Copernicus, and that the knowledge probably reached him through several intermediaries, but it seems unlikely that he developed these methods completely independently.

Legacies Today

Ancient Mesopotamia and the medieval Islamic world still influence the language of astronomy today. The division of a circle into 360° is taken directly from ancient Mesopotamian practice. Similarly, our division of time and angular units into successive sixtieths (minutes, seconds, etc) reflects the Mesopotamian sexagesimal number system that was then adopted by the ancient Greek astronomers and Islamic astronomers, and was eventually taken over in Europe. Another important astronomical concept used today that originated in ancient Mesopotamia is the zodiac, although it is probably now more relevant to astrology than to practical astronomy.

Many common astronomical terms are based upon Arabic words. Examples include altitude, azimuth, zenith and nadir, all of which are used in describing the positions of objects in the sky; almanac; and mathematical terms such as algebra and algorithm. Another legacy of Islamic astronomy is found in many common star names which are Latinized versions of the Arabic names: for example, Algol (β Per) comes from *al-ghul* ('the ghoul'), Rigel (β Ori) from *al-rijl* ('the foot [of Orion])', and Altair (α Aqu) from *al-ta'ir* ('the flying eagle'). These stars and more are listed in the following table.

Common Name	Star (Bayer name)	Arabic Name	Translation
Aldebaran	α Tau	al-dabaran	The follower
Algol	β Per	al-ghūl	The ghoul
Algorab	δ Crv	al-ghurab	The raven
Altair	α Aqu	al-tā'ir	The flying eagle
Deneb	α Cyg	danab al-dagāga	The tail of the hen
Mizar	ζ UMa	al-maraqq	The loins
Rigel	β Ori	al-rijl	The foot (of Orion)
Vega	α Lyr	al-wāqi'	The swooping (vulture)

Astronomical observations made by astronomers in the ancient and medieval periods still have an important role to play in modern science. For example, the gradual slowing down of the Earth's rate of rotation can be traced through the study of historical reports of lunar and solar eclipses. This slowing down is mainly caused by friction in the moon's tidal pull on the oceans, but there are other factors such as post-glacial uplifts and core-mantle coupling within the Earth. As a result the length of day is gradually increasing. Over long timescales the cumulative effect of the gradual lengthening of the day builds up to four or five hours by the time we get back to the ancient world. The effect of this 'clock error' (the difference between an ideal clock, for example an atomic clock, and using the daily rotation of the Earth as a clock) can be easily detected from even quite crude eclipse observations made in the ancient and medieval periods. In studying the changes in the Earth's rate of rotation and the geophysics that underlie them, scientists have made extensive use of eclipse observations recorded by Late Babylonian astronomers; Ibn Yūnus, al-Bīrūnī and other Islamic astronomers; and medieval Arabic chroniclers. Without the diligence of these early astronomers in recording their observations, it would not be possible to investigate this type of scientific problem.

The most important legacy left by Babylonian and Islamic astronomers, however, is the very idea of astronomy as a science. The Babylonians were the first people to undertake systematic astronomical observation and (equally importantly) to keep detailed and regular accounts of their observations. It was also in Babylon where mathematics was first applied to astronomy, an event that transformed the subject into an investigative science where abstract theories were proposed and astronomical phenomena could be predicted in advance using these theories. This initial step, the idea that phenomena can be modelled and predicted, is probably the biggest hurdle a science must overcome. Without Babylonian astronomy as a foundation, Greek astronomy would not have developed as it did – neither as quickly nor as successfully. And without the remarkable work by Islamic astronomers from the eighth to the fifteenth centuries to preserve and then dramatically develop Greek astronomy, knowledge of technical astronomy would have practically disappeared in Europe during the Middle Ages and there would have been no Copernicus, no Kepler and no Galileo in the history of astronomy. This history of astronomy is the history of a worldwide enterprise: only by looking beyond the familiar names in the history books can we begin to understand how we have got to where we are today.

Appendix

The Sexagesimal System

The sexagesimal (base-60) number system was developed in Mesopotamia and used by almost all astronomers from antiquity down to the Renaissance. Although strange to our eye, it is a very powerful number system in which it is possible to write very large numbers with comparatively few 'digits'.

The 'base' of a number system is the factor that each digit's 'place' is greater or smaller than a digit in the next place. For example, in our decimal (base-10) system, each place is greater or smaller than the next place by a factor of 10: hundreds, tens, units, tenths, hundredths, and so on. Moving a digit one place to the left increases its value by ten times and moving it to the right makes it ten times smaller. Number systems where this principle applies are called 'place-value' systems because the value of a digit depends upon its 'place' within the number.

The sexagesimal number system operates by exactly the same rules as our decimal system except that the base is 60 instead of 10 and we have 60 'digits' (0 to 59) rather than 10 (0 to 9 in the decimal system). In other words a factor of 60 is applied to a 'digit' moving one place within the number. Instead of hundreds, tens, units and tenths, we have three thousand six hundreds, sixties, units, and sixtieths.

Historians of astronomy conventionally write sexagesimal numbers by separating each place using a comma. A semicolon is used to indicate the transition from units to fractions (equivalent to the decimal point). Thus the number

2,43;30

means 2 sixties plus 43 units plus 30 sixtieths, or in decimal form 163.5.

Here are some sexagesimal numbers encountered in this book with their decimal equivalents:

Value	Sexagesimal form	Decimal form
One value of the solar velocity used in Babylonian System A lunar theory	28;7,30 degrees per month	28.125 degrees per month
Mean length of the synodic month according to the Babylonian System B lunar theory	29;31,50,8,20 days	29.530594135..... days
Number of months in the year according to the Babylonians	12;22,8 months	12.36888888..... months

Bibliography

General Works

Aaboe, A., *Episodes from the Early History of Astronomy*, Springer, 2001.

Evans, J., *The History and Practice of Ancient Astronomy*, Oxford University Press, 1998.

Neugebauer, O., *The Exact Sciences in Antiquity*, Brown University Press, 1957; reprint: Dover, 1967.

Neugebauer, O., *A History of Ancient Mathematical Astronomy*, Springer, 1975.

North, J., *The Fontana History of Astronomy and Cosmology*, Fontana, 1994.

Steele, J. M., *Observations and Predictions of Eclipse Times by Early Astronomers*, Kluwer, 2000.

Stephenson, F. R., *Historical Eclipses and Earth's Rotation*, Cambridge University Press, 1997.

Stephenson, F. R. and Green, D., *Historical Supernovae and their Remnants*, Oxford University Press, 2002.

Walker, C., ed., *Astronomy Before the Telescope*, British Museum Press, 1996.

Articles on many of the astronomers discussed in this book may be found in:

C. C. Gillespie (ed.), *Dictionary of Scientific Biography* (Charles Scribner and Sons) and H. Selin (ed.), *Encyclopaedia of the History of Science, Technology and Medicine in Non-Western Cultures* (Kluwer Academic Publishers).

Babylonian Astronomy

Aaboe, A., 'Scientific Astronomy in Antiquity', *Philosophical Transactions of the Royal Society of London* A 276, 1974, 21–42.

Brown, D., *Mesopotamia Planetary Astronomy-Astrology*, Styx, 2000.

Hunger, H., *Astrological Reports to Assyrian Kings*, Helsinki University Press, 1992.

Hunger H. and Pingree, D., *Astral Sciences in Mesopotamia*, Brill, 1999.

Neugebauer, O., *Astronomical Cuneiform Texts*, Lund Humphries, 1955; reprint: Springer, 1983.

Parpola, S., *Letters from Assyrian and Babylonian Scholars*, Helsinki University Press, 1992.

Rochberg, F., *Babylonian Horoscopes*, American Philosophical Society, 1998.

Rochberg, F., *The Heavenly Writing: Divination, Horoscopy, and Astronomy in Mesopotamian Culture*, Cambridge University Press, 2004.

Sachs, A. J., 'Babylonian Observational Astronomy', *Philosophical Transactions of the Royal Society of London* A 276, 1974, 43–50.

Sachs A. J. and Hunger, H., *Astronomical Diaries and Related Texts from Babylonia*, Österreichischen Akademie der Wissenschaften, 1988–; 5 volumes published to date.

Steele, J. M., 'Eclipse Prediction in Mesopotamia', *Archive for History of Exact Science* 54, 2000, 421–54.

Steele, J. M., 'A 3405: An Unusual Astronomical Text from Uruk', *Archive for History of Exact Science* 55, 2000, 103–5.

Koch-Westenholz, U., *Mesopotamian Astrology: An Introduction to Babylonian and Assyrian Celestial Divination*, Museum Tusculanum Press, 1995.

Greek Astronomy

Jones, A., *Astronomical Papyri from Oxyrhynchus*, American Philosophical Society, 1999.

Toomer, G. J., *Ptolemy's Almagest*, Duckworth, 1983; reprint: Princeton University Press, 1998.

Islamic Astronomy

Ali, J., *The Determination of the Coordinates of Cities: al-Bīrūnī's Tahdīd al-Amākin*, American University of Beirut, 1967.

Kennedy, E. S., *A Survey of Islamic Astronomical Tables*, American Philosophical Society, 1956.

Kennedy, E. S., *The Exhaustive Treatise on Shadows by al-Bīrūnī*, University of Aleppo, 1976.

Kennedy, E. S., *Studies in the Islamic Exact Sciences*, American University of Beirut, 1983.

King, D. A., *Astronomy in the Service of Islam*, Variorum, 1993.

King, D. A., *In Synchrony with the Heavens: Studies in Astronomical Timekeeping and Instrumentation in Medieval Islamic Civilization. Volume 1: The Call of the Muezzin*, Brill, 2004.

King, D. A., *In Synchrony with the Heavens: Studies in Astronomical Timekeeping and Instrumentation in Medieval Islamic Civilization. Volume 2: Instruments of Mass Calculation*, Brill, 2005.

Lorch, R., *Al-Farghānī: On the Astrolabe*, Franz Steiner Verlag, 2005.

Ragep, F. J., *Naṣīr al-Dīn al-Ṭūsī's Memoir on Astronomy*, Springer, 1993.

Rashed, R., ed., *Encyclopedia of the History of Arabic Science*, Routledge, 1996.

Said, S. S. and Stephenson, F. R., 'Solar and Lunar Eclipse Measurements by Medieval Muslim Astronomers', *Journal for the History of Astronomy* 27, 1996, 259–73 and 28, 1997, 29–48.

Said, S. S., Stephenson, F. R. and Rada, W., 'Records of Solar Eclipses in Arabic Chronicles', *Bulletin of the School of Oriental and African Studies* 52, 1989, 38–64.

Saliba, G., *A History of Arabic Astronomy: Planetary Theories During the Golden Age of Islam*, New York University Press, 1994.

Sayili, A., *The Observatory in Islam*, Türk Tarih Kurumu Basimevi, 1988.

Stephenson, F. R. and Said, S. S., 'Records of Lunar Eclipses in Arabic Chronicles', *Bulletin of the School of Oriental and African Studies* 60, 1997, 1–34.

Wright, R. R., *The Book of Instruction in the Elements of the Art of Astrology by al-Bīrūnī*, Luzan & Co., 1934.

Legacy of Babylonian and Islamic Astronomy

Kunitzsch, P. and Smart, T., *Dictionary of Modern Star Names: A Short Guide to Modern Star Names and Their Derivations*, Sky Publishing Corporation, 2006.

McCluskey, S. C., *Astronomies and Cultures in Early Medieval Europe*, Cambridge University Press, 1998.

Pingree, D., *From Astral Omens to Astrology – From Babylon to Bikaner*, Instituto Italiano per l'Africa e l'Oriente, 1997.

Swerdlow, N. M. and Neugebauer, O., *Mathematical Astronomy in Copernicus's De Revolutionibus*, Springer, 1984.

Author's Note

For convenience all dates in this book are given using the BC/AD system. Hijra dates may be obtained using the tables published by G. S. P. Freeman-Grenville, *The Islamic and Christian Calendars AD 622– 2222 (AH 1–1650)*, Garnet Publishing, 1995.

Ancient and medieval astronomers usually used the sexagesimal number system in their mathematics. An appendix explaining this system is provided, along with suggestions for further reading, at the end of this book.

The translations on the following pages have been taken from:

p. 24: S. Dalley, *Myths from Mesopotamia*, Oxford University Press, 1989, pp. 255–6.

p. 32: E. Reiner, *Babylonian Planetary Omens Part 4*, Brill, 2005, p. 41

p. 35: S. Parpola, *Letters from Assyrian and Babylonian Scholars*, Helsinki University Press, 1993, p. 107.

p. 36: S. Parpola, *Letters from Assyrian and Babylonian Scholars*, Helsinki University Press, 1993, pp. 231–4.

p. 43: A. J. Sachs and H. Hunger, *Astronomical Diaries and Related Texts from Babylonia. Volume III: Diaries from 164 B.C. to 61 B.C.*, Österreichische Akademie der Wissenschaften, 1996, p. 185.

p. 49: A. J. Sachs and H. Hunger, *Astronomical Diaries and Related Texts from*

Babylonia. Volume I: Diaries from 652 B.C. to 262 B.C., Österreichische Akademie der Wissenschaften, 1988, p. 49.

p. 67: A. J. Sachs and H. Hunger, *Astronomical Diaries and Related Texts from Babylonia. Volume I: Diaries from 652 B.C. to 262 B.C.*, Österreichische Akademie der Wissenschaften, 1988, p. 179.

p. 68: A. J. Sachs and H. Hunger, *Astronomical Diaries and Related Texts from Babylonia. Volume I: Diaries from 652 B.C. to 262 B.C.*, Österreichische Akademie der Wissenschaften, 1988, p. 207.

p. 89: D. A. King, *In Synchrony with the Heavens: Studies in Astronomical Timekeeping and Instrumentation in Medieval Islamic Civilization. Volume 1: The Call of the Muezzin*, Brill, 2004, p. 216.

p. 101: S. S. Said, F. R. Stephenson and W. Rada, 'Records of Solar Eclipses in Arabic Chronicles', *Bulletin of the School of Oriental and African Studies* 52, 1989, p. 47.

p. 101–2: S. S. Said, F. R. Stephenson and W. Rada, 'Records of Solar Eclipses in Arabic Chronicles', *Bulletin of the School of Oriental and African Studies* 52, 1989, p. 50.

p. 102: F. R. Stephenson and S. S. Said, 'Records of Lunar Eclipses in Arabic Chronicles', *Bulletin of the School of Oriental and African Studies* 60, 1997, p. 22.

p. 104–5: S. S. Said and F. R. Stephenson, 'Solar and Lunar Eclipse Measurements by Medieval Muslim Astronomers, I: Background', *Journal for the History of Astronomy* 27, 1996, p. 269.

In writing this book I have been fortunate to be able draw upon the work of a small group of historians of ancient and medieval astronomy whose groundbreaking work over the past 120 years has lead to the rediscovery of Babylonian and Islamic astronomical science. A few of these individuals, though by no means all, are mentioned in the text. Rather than cite all of their works using footnotes, which would take up half the book, references to several of the most important English-language publications are given in the bibliography.

Index